99-ECO-8UR

AF622614

# DEVELOPMENT OF A WATER QUALITY MODEL TO SUPPORT A PATHOGEN TMDL (NEWPORT BAY, CALIFORNIA)


by:

**John F. DeGeorge**
**Stacie Grinbergs**
Resource Management Associates, Inc.

**Jeffrey A. Soller**
**Adam W. Olivieri**
Eisenberg, Olivieri, & Associates, Inc.

**Joseph N. Eisenberg**
University of California, Berkeley


*2002*


Water Environment
Research Foundation

The Water Environment Research Foundation, a not-for profit organization, funds and manages water quality research for its subscribers through a diverse public-private partnership between municipal utilities, corporations, academia, industry, and the federal government. WERF subscribers include municipal and regional water and wastewater utilities, industrial corporations, environmental engineering firms, and others that share a commitment to cost-effective water quality solutions. WERF is dedicated to advancing science and technology addressing water quality issues as they impact water resources, the atmosphere, the lands, and quality of life.

For more information, contact:
Water Environment Research Foundation
601 Wythe Street
Alexandria, VA 22314-1994
Tel: (703) 684-2470
Fax: (703) 299-0742
www.werf.org
werf@werf.org

For non-subscriber sales information, contact:
IWA Publishing
Alliance House
12 Caxton Street, London SW1H 0QS,
United Kingdom
Tel: +44 (0)20 7654 5500
Fax: +44 (0)20 7654 5555
www.iwapublishing.com
publications@iwap.co.uk



Library of Congress Catalog Card Number: 2002114865

WERF ISBN 978-1-893664-68-5    IWAP ISBN: 978-1-84339-664-2



Irvine Ranch Water District

# ACKNOWLEDGMENTS

The project team acknowledges and thanks Drs. Robert C. Cooper, George Tchobanoglous, Jack Colford, and Edwin Lennette who served as the Health Advisory Committee for the Public Health Risk Assessment for the Newport Bay Watershed. The project team would also like to thank Drs. Roy Wolfe, April Garcia, Roger Fujioka, Sunny Jiang, and Stanley Grant who served as an independent review panel for the Newport Bay Watershed Committee, for their comments and suggestions. In addition to the Water Environment Research Foundation, the California Regional Water Quality Control Board and members of the Newport Bay Watershed Executive Committee provided funding for the Public Health Risk Assessment investigation, of which the work described herein was one component. The project team would also like to acknowledge and thank Ken Thompson, Debbi Clark, Larry Honeybourne, the Irvine Ranch Water District, and the Orange County Health Care Agency for their substantial contributions to this investigation.

## Report Preparation:

**Principal Investigator:**
James Hyde, B.S. (since January 2001)
Ken Thompson, B.S. (until December 2000)
*Irvine Ranch Water District*

**Project Team:**
John F. DeGeorge, Ph.D.
Stacie Grinbergs, M.S., P.E.
*Resource Management Associates*

Adam W. Olivieri, Dr. P.H., P.E.
Jeffrey A. Soller, M.S.
*EOA, Inc.*

Joseph N. Eisenberg, Ph.D.
*University of California, Berkeley*

Debbi Clark, M.S.
*Irvine Ranch Water District*

**Project Subcommittee**
Norman E. LeBlanc, Co-Chair
*Hampton Roads Sanitation District*

Philip B. Dorn, Ph.D., Co-Chair
*Shell Global Solutions US*

Fredric P. Andes
*Barnes & Thornburg*
Donald J. Brady, Ph.D.

*U.S. Environmental Protection Agency*

Betsy A. Cooper
*King County(Washington) Department of Natural Resources*

Robbin W. Finch
*Boise City (Idaho) Public Works*

Michael Haire
*U.S. Environmental Protection Agency*

John Morton
*ALCOA*

David Roberts (Served since March 2001)
*Washington Department of Ecology*

Jerald L. Schnoor, Ph.D. (Served until December 2001)
*University of Iowa*

Richard F. Schwer
*DuPont Company*

**Water Environment Research Foundation Staff:**

| | |
|---|---|
| Research Program Director: | Patricia J. Haddon |
| Project Manager: | Dean E. Carpenter, D.Env. |

# ABSTRACT AND BENEFITS

**Abstract:**

Newport Bay (Orange County, California) is listed by the California State Water Resources Control Board (SWRCB) as a water quality limited receiving water body because of sporadic exceedances of the fecal coliform water quality objectives for body contact recreation. Consistent with federal and state requirements, a Total Maximum Daily Load (TMDL) is being implemented in the watershed. The fecal coliform TMDL in Newport Bay is a phased approach for understanding and controlling the microbiological water quality in the Bay to ensure the reasonable protection of the Bay's beneficial uses. An important initial step within that TMDL was to assess the impairment of the body contact recreation beneficial use (REC-1) of Newport Bay receiving waters through characterizing the risk of illness associated with REC-1 exposure. A health risk assessment investigation, developed to characterize that risk, involved the integration of a population based model of disease transmission, a water quality modeling component necessary for estimating pathogen dose as part of the exposure assessment, and site-specific population use and receiving water data collection.

The Water Environment Research Foundation funded the water quality modeling component of the health risk assessment investigation, which is the focus of this report. Also provided within this report is an overview of the health risk assessment methodology, a summary of the major findings from the risk assessment investigation, and a discussion of how the health risk methodology may be applied to other watersheds where impairment of the REC-1 beneficial use is in question.

The major findings of the health risk assessment investigation indicated that (1) the risk of illness from REC-1 use in Newport Bay, estimated using two separate methods was generally below levels considered tolerable by US EPA, and (2) the reduction of controllable sources of pollution would not appreciably reduce the existing risk. Based on the collection of site-specific exposure data and the health risk characterization, it was determined that evaluating the impairment of the REC-1 beneficial use requires a more rigorous and comprehensive health based approach than that prescribed by the current regulations for recreational waters.

**Benefits:**

- Summarizes a method that may be used to interpret a fecal coliform TMDL in terms of beneficial use impairment.
- Presents beneficial use impairment analysis and findings in a health-based framework that can guide the technical and management review and analysis of the problem characterization component of a TMDL in a defensible manner.
- Provides water quality and local public agency managers an approach to balance the benefits of imposing additional controls to improve water quality and the estimated costs of these controls.
- Discusses the integration of water quality modeling with microbial risk assessment to facilitate TMDL interpretation with respect to beneficial use impairment.
- Describes fecal coliform and coliphage water quality modeling of Newport Bay.
- Summarizes the theoretical use of alternative management scenarios to explore potential control measures.

**Keywords:** microbial risk assessment, beneficial use impairment

# TABLE OF CONTENTS

# LIST OF TABLES

# LIST OF FIGURES

# EXECUTIVE SUMMARY

## ES.1 Background

The California Regional Water Quality Control Board, Santa Ana Region (RWQCB), identified and listed Newport Bay as a water quality limited receiving water body in accordance with Section 303(d) of the Clean Water Act. This designation indicates that applicable water quality standards are not being attained or expected to be attained with the implementation of technology-based controls. The 303(d) List identifies pathogens as a stressor of water quality impairment and indicates that urban runoff and/or storm sewers are a primary source of the pathogens. On April 9, 1998, the RWQCB adopted a Total Maximum Daily Load (TMDL) for fecal coliform in Newport Bay.

The RWQCB fecal coliform TMDL is based on a phased approach for understanding and controlling the microbiological quality in Newport Bay and ensuring that the beneficial uses in the Bay are protected. The complexity of this fecal coliform TMDL derives primarily in interpreting the following question: "Do occasional exceedances of the fecal coliform objective for body contact recreation necessarily imply an impairment to the beneficial use?" Several issues make answering this question difficult: (1) The risk to human health from contact recreation stems primarily from exposure to human viruses, yet the water quality objectives are based on bacterial indicator organisms; (2) The linkage between the fecal coliform concentrations and adverse health effects is weak, and the improvement provided by other bacterial indicators relative to virus concentrations is debatable; (3) There are significant temporal and spatial variations in water quality and in recreational use in Newport Bay; (4) The low number of people recreating in the Bay at most locations make an epidemiological study impractical; and (5) Limited data are available to identify controllable sources of coliform and to evaluate the potential human health related benefits by reducing coliform loading from those sources.

## ES.2 Purpose

A health risk assessment investigation was initiated at the onset of the TMDL to provide technical data that would allow the RWQCB to characterize, in a technically defensible manner, the impairment of the body contact recreation (REC-1) beneficial use in Newport Bay, and to evaluate "reasonable" protection of the use consistent with California State water quality regulations[1]. The water quality modeling component which was used to estimate pathogen dose as part of the exposure assessment, was funded by the Water Environment Research Foundation and is the primary focus of this report.

The objectives of the health risk assessment investigation were: (1) To characterize the existing relative risk to public health posed by exposure to pathogens derived from human sources via recreational contact in Newport Bay; (2) To provide estimates of the relative risk associated with alternative levels of exposure and/or water quality (due to structural or programmatic changes), and to compare those results with the results representing existing conditions; and (3) To integrate the results of the risk assessment with planning level costs for implementing control alternatives that may lead to water quality improvements and reductions in public health risk. The goal of the water quality modeling component of the health risk assessment investigation was to extend the existing numerical modeling capability for Newport Bay to include simulation of fecal coliform and coliphage in support of the fecal coliform TMDL for the Newport Bay watershed. The numerical model provides temporally and spatially varying concentrations of fecal coliform and coliphage for use by the health effects model. Specific objectives of the water quality modeling were to:

- Configure and calibrate a fecal coliform and coliphage transport model based on the existing Research Management Associates (RMA) model of Upper Newport Bay;
- Perform uncertainty analysis on model results;
- Develop the a numeric linkage between the RMA model and the disease transmission model;
- Perform simulations for a set of alternative control strategies; and
- Provide the resulting coliphage concentration results to the health effects model.

## ES.3 Methods

The health risk assessment approach was to take advantage of the available infectious disease and dose-response data to develop a dynamic population based mathematical model that characterizes the human disease risk of waterborne pathogen exposure. This approach makes explicit the mechanistic aspects of the infectious disease process and provides a transparent structure for clinical data to be incorporated into the mathematical model. The health risk model emphasizes the importance of how the epidemiological status (susceptible, infected, diseased, and immune) of individuals within a population varies over time, implicitly accounting for person-to-person spread of disease. The risk assessment methodology is consistent with the 1996 EPA framework for microbial risk assessment and further extends work previously carried out in the microbial risk field by characterizing the population based risk from recreational activity in a metric that is comparable to U.S. EPA water quality criteria endpoints, specified as risk to individual swimmers. A schematic diagram of the investigation is presented in Figure ES.1. As shown, the investigation included site-specific data collection efforts, water quality modeling, and a disease transmission modeling component.

---

[1] California Water Code, Section 13241.

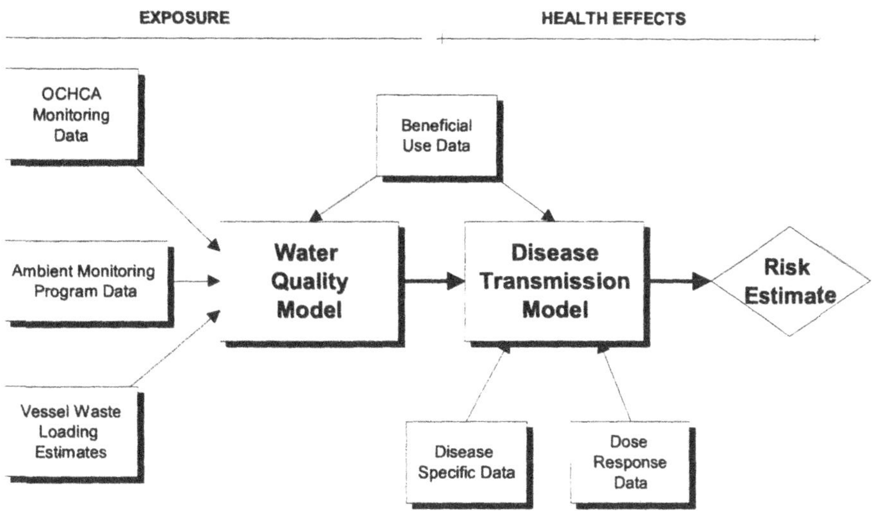


**Figure ES-1. Schematic Diagram of Health Risk Assessment Investigation**

## ES.4 Health Risk Investigation Findings

### ES.4.1 303(d) Listing of Newport Bay for Fecal Coliform.

Exceedances of the Basin Plan fecal coliform objectives for the REC-1 beneficial use are temporally sporadic, geographically limited, and generally occur during the time of the year when REC-1 use is low and/or in areas of the Bay where the level of body contact recreation is low or is prohibited.

### ES.4.2 Relative Risk to Public Health from REC-1 Contact in Newport Bay

The health risk assessment was carried out employing a number of health protective assumptions (for example, the model organism was assumed to be as infectious as rotavirus and as prevalent and persistent in the environment as male specific coliphage; boundary conditions in the water quality model were developed based on the maximum concentrations observed; and all bacteriological observations reported below the detection limits were assumed to correspond to concentrations at the detection limit). Given the conservative nature of the assumptions that were employed, it is extremely likely that the estimated risk levels presented herein are higher than the "actual" levels of risk encountered by the population. Nevertheless, the risk of illness from REC-1 use in Newport Bay was estimated using two separate methods (i.e. use of disease transmission modeling and use of US EPAs relation between enterococcus and health risk), both of which found that the estimated risk from REC-1 use was generally below levels considered tolerable by US EPA.

### ES.4.3 Controllable Sources

Urban runoff and/or storm sewers are identified in the 303(d) listing as the likely source of pathogens in Newport Bay. However, the results of the health risk assessment indicate that neither of those sources substantially impact the risk to public health from body contact recreation in Newport Bay during the time of the year that the vast majority of the recreation occurs. Viral loadings from vessel sanitary waste and from swimmers were also investigated as two potential sources of pathogen loading in Newport Bay. Based on the estimated costs to implement a public outreach program and the estimated benefit in terms of reducing enteric virus disease, the reduction of vessel sanitary waste may be a control measure worth further consideration if risk managers feel that reduction of public health risk below the current levels is warranted.

### ES.4.5 Interpretation of the Basin Plan Fecal Coliform Objective

Based on the observed patterns and levels of recreational use in Newport Bay and the results of the health risk assessment investigation, it is clear that interpreting the impairment of the REC-1 beneficial use in Newport Bay requires a more rigorous and comprehensive health based evaluation than the pass/fail test (i.e. simple comparison of water quality indicator data against water quality objectives / standards regardless of exposure conditions and sources) that has evolved to regulate the quality of recreational waters.

A more comprehensive approach is consistent with a health based monitoring approach for recreational waters recently outlined by the World Health Organization (WHO 1999) in which experts called for "an improved approach to the regulation of recreational water that better reflects health risk and provides enhanced scope for effective management intervention". In a similar manner, EPA's strategy for setting water quality criteria was to vary the maximum allowable density based on the level of recreational use that a particular site receives.

In this investigation, a risk assessment approach was taken to address the question of beneficial use impairment. If sufficient site-specific data or knowledge are available however, a less comprehensive approach may be suitable to address the issue of use impairment. For example, data from a sanitary survey, relevant water quality data, or specific knowledge about microbiological sources may allow a more qualitative approach to be employed. A site by site summary is provided in Table ES.1 demonstrating how such an approach may be used to facilitate interpretation of the Basin Plan Objective for fecal coliform with respect to risk to public health, and thus the impairment of the beneficial use. The summary provided in Table ES.1 uses observed fecal coliform concentrations, levels of exposure, and seasonality to derive a relative level of public health concern associated with each of the monitoring sites discussed in the report.

**Table ES.1 Summary of Relative Public Health Concern Related to Fecal Coliform Exceedances in Newport Bay**

| Site Name | Recreational Use Level [1] | Exceedances of FC WQC During Dry Season [2] Log Mean | 90th %ile | Level of Public Health Concern Dry Season [7] | Exceedances of FC WQC During Wet Season [3] Log Mean | 90th %ile | Level of Public Health Concern Wet Season [7] | Potential Future Controls [4] Vessel Waste | Swimmer Loading | Tributary Inflow | Others [9] |
|---|---|---|---|---|---|---|---|---|---|---|---|
| Vaughn's Launch | Use Prohibited | | √ | Very Low | √ | √ | Low | - | - | - | - |
| Ski Zone | | | | Very Low | | √ | Very Low | - | - | - | - |
| Big Canyon Creek | | | √ | Very Low | √ | √ | Low | - | - | - | - |
| San Diego Creek - Campus Dr. | | | √ | Very Low | √ | √ | Low | - | - | - | - |
| Backbay Drive Pipe [5] | | | √ | Very Low | | √ | Very Low | - | - | - | - |
| Santa Ana Delhi Channel | | √ | √ | Low | √ | √ | Low | - | - | - | - |
| 43rd Street Beach | Low Use | √ | √ | Medium | | √ | Low | √ | - | - | Flushing |
| 38th Street Beach | | | √ | Low | | √ | Low | - | - | - | Flushing |
| 33rd Street Channel | | | √ | Low | | √ | Low | - | - | - | Flushing |
| Rhine Channel | | | | Very Low | | √ | Low | - | - | - | Flushing |
| Sapphire Avenue Beach | | | | Very Low | | | Very Low | - | - | - | - |
| N Street Beach | | | | Very Low | | √ | Low | - | - | - | - |
| Promontory Point Channel | | | | Very Low | | | Very Low | - | - | - | - |
| De Anza | | | | Very Low | | √ | Low | - | - | - | - |
| Grand Canal | | | | Very Low | | | Very Low | - | - | - | - |
| Newport Blvd. Bridge | | | √ | Low | | √ | Low | - | - | - | Flushing |
| Park Avenue Beach | Medium Use | | | Very Low | | | Very Low | - | - | - | - |
| Onyx Avenue Beach | | | | Very Low | | √ | Low | - | - | - | - |
| Ruby Avenue Beach | | | | Very Low | | √ | Low | - | - | - | - |
| Bayshore Beach [6] | | | √ | Medium | | √ | Low | √ | - | - | - |
| Via Genoa Beach | | | | Very Low | | √ | Low | - | - | - | - |
| 15th Street Beach | | | √ | Low | | √ | Low | - | - | - | - |
| 10th Street Beach | | | | Very Low | | √ | Low | - | - | - | - |
| Alvarado/ Bay Isle Beach | | | | Very Low | | √ | Low | - | - | - | - |
| Abalone Avenue Beach | | | | Very Low | | √ | Low | - | - | - | - |
| Garnet Avenue Beach | | | | Very Low | | √ | Low | - | - | - | - |
| Harbor Patrol Beach | | | | Very Low | | √ | Low | - | - | - | - |
| 19th Street Beach | High Use | | | Very Low | | √ | Low | - | - | - | - |
| Rocky Point Beach | | | | Very Low | | | Very Low | - | - | - | - |
| North Star Beach | | | | Very Low | | √ | Low | - | - | - | - |
| Lido Yacht Club Beach | | | | Very Low | | √ | Low | - | - | - | - |
| Newport Dunes - Middle | Extra High Use | | | Low | | √ | Medium | √ | √ | - | Flushing |
| Newport Dunes - West | | | | Low | | √ | Medium | √ | √ | - | Flushing |
| Newport Dunes - North [8] | | | √ | High | | √ | Medium | √ | √ | - | Flushing |
| Newport Dunes - East | | | | Low | | √ | Medium | √ | √ | - | Flushing |

1. Use Levels are based on data collected as part of this investigatic
2. Dry Season defined as May through September
3. Wet Season defined as October through April
4. Potential structural changes or management options
5. Dry weather flow is now diverted to sanitary sewer
6. High Use on holidays
7. Level of Public Health Concern is derived as shown to the right
8. Exceedances may have been linked to flow from the Backbay Drive Pipe, see note 5.
9. Other potential control measures not investigated as part of the HRA

Level of Public Health Concern
Dry Season

| Indicator Concentration | Exposure: None | Low | Medium | High | Extra high |
|---|---|---|---|---|---|
| Low | Very Low | Very Low | Very Low | Very Low | Low |
| Medium | Very Low | Low | Low | Medium | High |
| High | Low | Medium | Medium | High | High |

Level of Public Health Concern
Wet Season

| Indicator Concentration | Exposure: None | Low | Medium | High | Extra high |
|---|---|---|---|---|---|
| Low | Very Low | Very Low | Very Low | Very Low | Low |
| Medium | Very Low | Low | Low | Low | Medium |
| High | Low | Low | Low | Medium | High |

Note: The relative level of Public Health Concern shown in these tables is based on the combination of exposure level, indicator concentration, basic principles of sanitary engineering, and professional judgement.

Risk managers in other watersheds may use a similar approach to that shown in Table ES.1 to consider the potential impairment of the REC-1 beneficial use within their watersheds. By considering potential levels of exposure and seasonality in conjunction with water quality data, the reasonable protection of the REC-1 beneficial use may be addressed in a streamlined manner emphasizing public health protection while concurrently balancing the costs of additional control measures.

CHAPTER 1.0

# INTRODUCTION TO THE HEALTH RISK ASSESSMENT

## 1.1 Background

The Santa Ana Region California Regional Water Quality Control Board (RWQCB) identified Newport Bay (Orange County, California) as a water quality limited receiving water body. This designation indicates that applicable water quality standards were not being attained or expected to be attained with the implementation of technology-based controls. The Bay is also included on the 1998 California 303(d) List, which identifies pathogens as a stressor of water quality impairment. Under these conditions, the Clean Water Act requires the development of a Total Maximum Daily Load (TMDL), a method for implementing water quality standards based on the relation between sources of pollutants and in-stream water quality conditions. However, in California, a TMDL also must be conducted within the context of the Porter-Cologne Act (California Water Code), which requires that the potential costs associated with regulatory actions be considered in relation to the benefits projected for those actions. Consistent with both federal and state requirements, the Newport Bay Watershed is carrying out a TMDL for fecal coliform bacteria.

The Newport Bay fecal coliform TMDL (CA RWQCB 1999) is a phased approach for understanding and controlling the microbiological water quality in the Bay and focuses on generating information to ensure the reasonable protection of the Bay's beneficial uses. The fundamental question motivating investigations during the first phase of the TMDL is: "Do occasional exceedances of the fecal coliform objectives for body contact recreation necessarily imply an impairment to the beneficial use?"

Several issues make answering this question difficult. The risk to human health from contact recreation stems primarily from exposure to human viruses (Cabelli 1983; Fankhauser et al. 1998; Levine and Stephenson 1990; Mead et al. 1999; Palmeteer et al. 1991; Sobsey et al. 1995; WHO 1999), yet the water quality objectives are based on bacterial indicator organisms. The linkage between the fecal coliform concentrations and adverse health effects is weak (Gerba et al. 1979; Kay et al. 1994; McLaughlin and Rose 2000), and the improvement provided by other bacterial indicators relative to virus concentrations is debatable (Fleisher 1991; Fleisher et al. 1993; Gantzer et al. 2000; Sobsey et al. 1995). There are significant temporal and spatial variations in water quality and in recreational use in Newport Bay. The low number of people recreating in the Bay at most locations makes an epidemiological study impractical. Limited data were available to identify controllable sources of coliform and to evaluate the potential human health related benefits by reducing coliform loading from those sources.

A health risk assessment was initiated at the onset of the TMDL to provide data that would allow the RWQCB to characterize, in a technically defensible manner, the impairment of the body contact recreation (REC-1) beneficial use in Newport Bay, and to evaluate "reasonable" protection of the use consistent with California State water quality regulations. The objectives of the health risk assessment were to:

- Conservatively characterize the existing relative risk to public health posed by exposure to enteric viruses via recreational contact in Newport Bay;
- Provide estimates of the relative risk associated with alternative levels of exposure and/or water quality (due to structural or programmatic changes), and to compare those results with the results representing existing conditions; and
- Integrate the results of the health risk assessment with planning level cost estimates for the alternative exposure levels investigated.

## 1.2 Methods

The fundamental approach was to develop a mathematical model of disease transmission to conservatively characterize the human disease risk from waterborne pathogen exposure in Newport Bay. Existing conditions in the Bay were simulated and estimates of disease risk were generated. Similarly, potential structural modifications (i.e., treatment of stream flow and diversion of stormwater) and programmatic changes (i.e., public information campaigns) were identified, characterized in terms of planning level costs, and simulated. The existing risk in the Bay was then compared to levels of risk considered tolerable by EPA, and the potential effectiveness of the alternatives was characterized and presented with planning level costs.

The approach builds on previously published work (Eisenberg et al. 1996; Eisenberg et al. 1998; EOA Inc. 1995; EOA Inc. et al. 1996; EOA Inc. and U.C. Berkeley School of Public Health 1995; EOA Inc. and U.C. Berkeley School of Public Health 1999), and is consistent with the 1996 EPA framework for microbial risk assessment (International Life Sciences Institute - Risk Science Institute 1996) and EPA's guidance for pathogen TMDLs (U.S. EPA 2001). The risk characterization methodology employed extends previous population-based microbial risk assessment work by characterizing the population-based risk from recreational activity in a metric that is comparable to EPA water quality criteria endpoints (U.S. EPA 1986; U.S. EPA 2000), specified as risk to individual swimmers.

The methodology employed explicitly models the mechanistic aspects of the infectious disease process from a population-based perspective and provides a transparent structure for clinical, dose-response, and exposure data to be incorporated into the modeling process. Fundamental to the investigation were data collection efforts, water quality modeling, and disease transmission modeling (Figure 1-1). Data that were generated and/or collected for this investigation include: (1) Estimates of the frequency and duration of recreational use of Newport Bay; (2) An estimate of virus and coliform loading to the Bay from bathers; (3) A characterization of the size of the population that is most likely to recreate in Newport Bay; (4) An estimate of virus and fecal coliform loading to the Bay from vessels; (5) A data collection effort augmenting the existing bacteriological database for the Bay; and (6) A comprehensive review of the literature related to indicator organisms used for characterizing microbial water quality and microbial risk assessment.

### 1.2.1 Health Protective Assumptions

In this investigation, a combination of conservative (health protective) and pragmatic assumptions was employed so that the result of the assessment could reasonably be considered health protective. This approach is consistent with previously published work (Haas et al. 1996) and is a commonly applied approach in public health based risk assessment.

The most important health protective assumptions in this investigation included: (1) use of a model organism that was assumed to be as infectious as rotavirus; (2) use of a model organism as prevalent and persistent in the environment as male specific coliphage; (3) use of the boundary conditions in the water quality modeling that were based on the maximum concentrations observed; and (4) all microbiological observations reported below the detectable limits were assumed to correspond to concentrations at the detection limit.

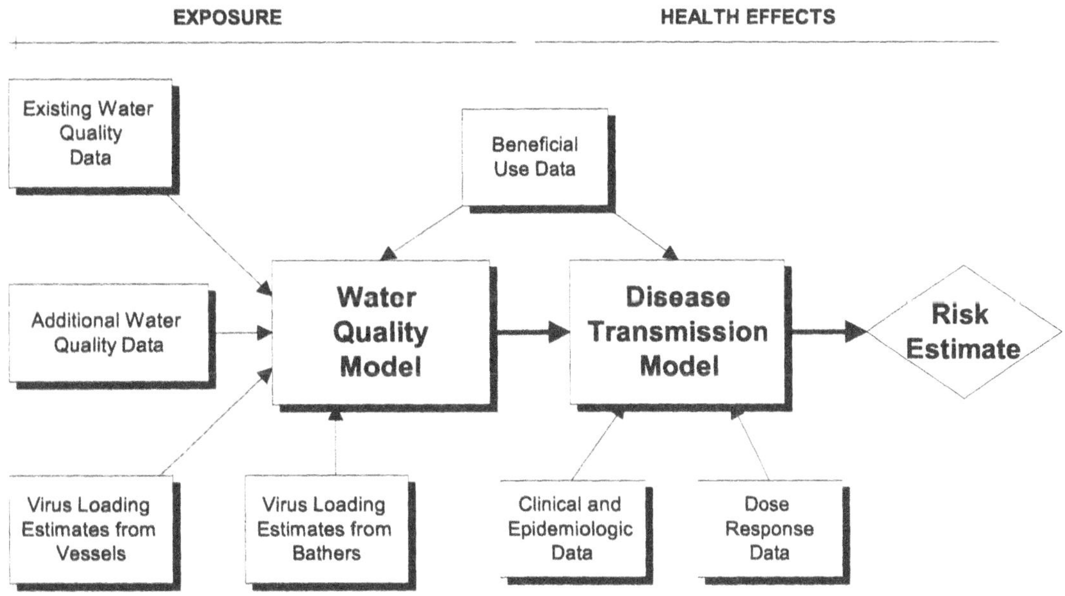


**Figure 1-1 Fundamental Components of the Health Risk Characterization**

### 1.2.2 Link Between Water Quality Data and Risk to Public Health

A critical aspect of this investigation was to identify an appropriate link between water quality data andrisk to public health. Based on research conducted over the last twenty years, it is very likely that the primary risk associated with recreational exposure to waterborne pathogens is from viral contaminants (Cabelli 1983; Fankhauser et al. 1998; Levine and Stephenson 1990; Mead et al. 1999; Palmeteer et al. 1991; Sobsey et al. 1995; WHO 1999). The local water quality objectives for contact recreation, however, are delineated in terms of bacterial indicator organism concentrations, and a substantial database of total coliform, fecal coliform, and enterococcus data were available from the Orange County Health Care Agency (OCHCA). The utility of bacterial organisms as indicators for viral pathogens in ambient waters has been widely debated (Fleisher 1991; Fleisher et al. 1993; Gantzer et al. 2000; Gerba et al. 1979; Kay et al. 1994; McLaughlin and Rose 2000; Sobsey et al. 1995). An alternative indicator organism for this investigation was therefore sought.

EPA's Protocol for Developing Pathogen TMDLs (U.S. EPA 2001) recommends that in the selection of an indicator organism consideration be given to established water quality standards, alternative indicators, and designated beneficial uses. Based on a detailed review of the literature, male specific coliphages were selected as the indicator of enteric viruses for this investigation. According to Sobsey and co-workers "These coliphages are a group of small icosahedral bacteriophages infecting male strains of *E. Coli*, that are physically and chemically related more closely to human enteric viruses" (Sobsey et al. 1995) than bacterial indicators. It has also been reported that the survival of male specific coliphages more closely resembles that of viruses as compared to bacterial indicators (Armon and Kott 1996; Baldini et al. 1978; Borrego et al. 1987; Griffin et al. 1999; IAWPRC Study Group 1991; McLaughlin and Rose 2000; Paul et al. 1997; Sinton et al. 1999; Wommack et al. 1996).

The use of male specific coliphage as a viral indicator began to appear in the literature in the 1980s as researchers reported that these organisms may fulfill many of the essential requirements of a viral indicator (Havelaar 1987; Havelaar et al. 1986; Havelaar et al. 1984; Havelaar et al. 1993; IAWPRC Study Group 1991; Paul et al. 1997; Sobsey et al. 1995). Discussion on the appropriate application of these organisms as viral indicators has continued through the current time in the literature. Recently, Grabow and co-workers published an update on bacteriophages as models for viruses in water and report that "male specific coliphages have been found to outnumber cytopathogenic enteric viruses by a factor of about 100. This implies that the absence of (male specific) phages from raw and treated water supplies offers a meaningful indication of the absence of human enteric viruses" (Grabow 2001). The use of phages as models/surrogates for enteric viruses is "based particularly on structure, composition, size, and mode of replication which resemble enteric viruses much closer than commonly used bacterial indicator of fecal pollution such as coliforms and enterococci.... In addition, phages closely meet the requirements of models/surrogates for enteric viruses, survive longer in natural waters than enteric viruses, and ... fail to multiply in the environment" (Grabow 2001).

A close review of the literature also indicates that male specific coliphages are present in water in equal or higher numbers compared to enteric viruses. Studies illustrating this point have been reported by (Havelaar et al. 1993), (Gold et al. 1992), (AWWARF 1999), (Dutka et al. 1987), (Kott et al. 1974), (Stetler 1984), (Morinigo et al. 1992), and (Jagals et al. 1995). In these studies, the ratio of coliphage to viruses varied significantly, both between and within investigations. However, in all of these studies coliphages were found in equal or greater numbers than viruses. The ratio of coliphage to viruses reported in those investigations ranged from ~1:1 to ~1,000,000:1, with levels most commonly reported between 100:1 and 1000:1.

A ratio of 1:1 (coliphage:enteric virus) was selected for this investigation to maintain the conservative nature of the assessment. By assuming that each coliphage enumerated represents a viable virus, the investigation assumed the "worst case" scenario reported in the literature. Therefore, the approach employed in this investigation most likely overestimates the actual presence of viruses of more public health concern.

### 1.2.3 Water Quality Monitoring

Within the Newport Bay watershed there are a number of potential sources of microbiological pollution. Based on staff reports from the RWQCB (CA RWQCB 1999), those sources include tributary inflows (San Diego Creek, Santa Ana Delhi, other storm drains in Upper Newport Bay, and Back Bay Drain), food wastes, discharge of sanitary vessel waste, and wildlife fecal waste including waterfowl that inhabit the Bay and its environs. Further, additional potential sources of microbiological loading to Newport Bay include leakage of sewer lines (Metcalf and Eddy 1981), recreators (EOA Inc. 1995; Yates et al. 1997), animal wastes from domestic animals (Young and Thackston 1999), and illegal and/or illicit waste discharges from industrial, commercial, and/or residential sources.

#### 1.2.3.1 Routine Monitoring.

The OCHCA has monitored Newport Bay for total and fecal coliform on (approximately) a weekly basis over approximately the last 30 years. From the beginning of their monitoring effort through approximately 1998, total and fecal coliform data were collected. Starting in 1999 and continuing through the present, total coliform, *E. Coli*, and enterococcus data have been collected. Between 1990 and 2000, a total of approximately 26,625 analyses for total coliform, fecal coliform, *E. Coli*, and enterococcus were analyzed by OCHCA for approximately 34 Newport Bay sites. A review of these data indicate the following general observations (EOA Inc. et al. 2001):

- Most sites (27/35) have wet and dry season log mean fecal coliform values less than the 200 MPN/100mL water quality objective;
- Most sites (24/35) had wet season 90$^{th}$ percentile values of the reported fecal coliform observations above the 400 MPN/100mL water quality objective;
- Approximately one half (17/35) of the sites had dry season 90$^{th}$ percentile values above the 400 MPN/100mL water quality objective;
- Of the 12 sites with 1999 and 2000 fecal coliform 90$^{th}$ percentile values above 400 MPN/100mL, four are located in areas where recreational activities are prohibited (Santa Ana Delhi Channel, San Diego Creek @ Campus, Big Canyon Wash, and Vaughn's Launch), and three are near areas which receive substantial recreational use (15$^{th}$ Street, Bayshore Beach, and Newport Dunes North). The other sites are areas that are used infrequently for recreational activity; and
- The percent of sites with log mean enterococcus concentrations above EPA's marine criterion of 35/100mL was higher than the percent of sites above the fecal coliform log mean objective for the time period of 1999 – 2000 for both wet (13/35 compared to 4/35) and dry (6/35 compared to 2/35) periods.

#### 1.2.3.2 Site-Specific Monitoring for Health Risk Assessment.

A three component ambient monitoring program was developed to augment the OCHCA bacteriological database with additional data to support the water quality monitoring efforts for the health risk assessment. The first component was a screening phase in which male specific coliphage monitoring was amended to the existing OCHCA monitoring program. The purpose of this component was to establish a preliminary set of data, covering Newport Bay and tributaries, that could be used to assess the relative levels of indicator organisms present. Approximately 135 water samples were collected and analyzed for male specific coliphage. The analytical results indicate that 117 of 135 analyses were below the detection limit of 1 pfu/100mL.

During the second component of the ambient monitoring program, water samples were collected to establish boundary condition data and in-Bay coliphage data for calibration of the water quality model. A total of 260 samples were collected during this component, and were analyzed for *E. coli*, fecal coliform, total coliform, enterococcus, male specific coliphage, electrical conductivity, flow (major inflows and storm drains), salinity, and temperature. Samples were collected at three distinct types of sites: Upper Bay inflows, Lower Bay inflows, and in-Bay stations.

The final component of the ambient monitoring program was to investigate the change in indicator organism distribution throughout the course of a tidal cycle. To achieve this goal, total coliform, fecal coliform, male specific coliphage, and enterococcus samples were analyzed on two days throughout the course of a full tidal cycle. The monitoring included slack after flood and slack after ebb samples. On each day, eight sites were monitored no more than three hours apart throughout the tidal cycle. Results from all three components of the ambient monitoring program were used during the water quality modeling component of the investigation.

### 1.2.4 Beneficial Use Assessment

A beneficial use assessment program was designed and carried out to describe qualitatively and quantitatively the population that may be exposed to pathogenic microorganisms through body contact recreation in Newport Bay. The beneficial use assessment program was composed of two parts: beneficial use monitoring (counting number of recreators) and beneficial use surveying (asking recreators a series of questions). The beneficial use assessment was based on a purposeful sampling design in which recreational use in the Bay was monitored during 36 days. Monitoring was based on a randomized sampling plan stratified by season and type of day (weekday, weekend, holiday), and was purposely weighted to emphasize data collection during the summer (high use) period. The number of recreators at representative recreational sites was documented during the summer, fall, winter, and spring seasons. Beneficial use monitoring sites were selected based on those sites currently monitored by the OCHCA and on discussions with local stakeholders. Each of the sites was categorized by level of use (use prohibited, low, medium, high, extra high). It was assumed that each of the sites categorized as a particular use level received similar recreational activity. Representative sites for each use level were selected for beneficial use monitoring and surveying.

Information on recreating behavior was used to develop an estimate of virus and coliform loading to the Bay from bathers. Loading to the Bay from bathers was estimated based on the percentage of recreators swimming during a given interval (from survey responses), the expected number of recreators at a given time (from beneficial use assessment monitoring), the expected duration distribution for swim events (from survey data), and values for virus and coliform loading per time swimming (from literature). Loading to the Bay from bathers was used during the water quality modeling component of the investigation.

Similar to the beneficial use assessment, an initial estimate of enteric virus and fecal coliform loading to Newport Bay from vessels was also developed. Information used to estimate the loadings was collected from various sources including the City of Newport Beach Marine Department, interviews with representatives from private and public harbors in Newport Bay, the Orange County Harbor Master and Sheriff, a survey of boaters conducted specifically for this investigation, discussions with boat cleaners and holding tank pumpout services in Newport Bay, and aerial photographs.

Fecal coliform and viral loading to the Bay from vessel sanitary waste was computed based on a combination of site-specific and literature based values. For example, for viral loading the data required are the number of boats resident in the bay (4250), the estimated (from survey results) percent of boats that discharge waste to the Bay (10%), the person hours per boat per day (2.9 for weekdays, 11.4 for weekends), the grams of stool per capita per day (150 grams stool/day per capita) (Feachem et al. 1983), the number of infective virus particles per gram of stool ($10^8$ infective rotavirus particles per gram stool from infectious persons) (Flewett and Woode 1978), and the percent of the population shedding viruses at a given time. Loading to the Bay from vessels was also used during the water quality modeling component of the investigation.

### 1.2.5 Water Quality Modeling

Water quality modeling was used in the exposure assessment component of this investigation to characterize the levels of pathogens present in Newport Bay. The purpose of the water quality modeling was to provide temporally and spatially varying concentrations of coliphage to the health effects (disease transmission) model. These temporally and spatially varying concentrations of coliphage are used in the disease transmission model in conjunction with site-specific patterns of beneficial use in Newport Bay to define exposure from recreational contact. A second goal of the water quality modeling was to evaluate alternative control strategies that may affect viral loading. The water quality modeling is discussed in detail in Chapters 2 through 7.

### 1.2.6 Conceptual Model for Disease Transmission

Based on available dose-response and clinical data, and on relative public health importance of pathogens for which sufficient data are available, rotavirus was selected as the representative virus for enteric virus infections and disease (Black et al. 1984; Champsaur et al. 1983; Gurwith et al. 1981; Kapikian et al. 1980; Mead et al. 1999). Rotavirus is the most infective virus for which published dose-response studies are available (Haas et al. 1999), and also exhibits several critical epidemiological properties associated with other enteric viruses, such as the potential for person-to-person transmission of infection and (short term) protection from reinfection.

A conceptual model for disease transmission was developed (EOA Inc. et al. 2001) emphasizing the epidemiologic status of individuals within a population, and how that status varies over time (Figure 1-2). The model also explicitly accounts for person-to-person transmission of disease. State variables in the disease transmission model are used to track the number of people that are in each of the epidemiologic states at any point in time, and the sum of the state variables equals the total population (N). Rate parameters determine the movement of the population from one state to another. The rate parameters include $\beta$ (rate of acquiring infection), σ (rate of recovery from infectious states), and γ (rate of decline in immunity). Rate parameters are determined through literature review or are functions of model parameters, which are determined through literature review.

To illustrate how the model tracks movement of population members between states, consider State S, the susceptible portion of the population. At any given point in time, three processes affect the number of susceptible individuals within the population: (1) some members of the population will move from state S to state C, the carrier state (this happens at rate $\beta_{SC}$); (2) some members will move from state S to state D, the diseased state (this happens at rate $\beta_{SD}$); and (3) other members of the population will move from state P, the post-infection state back to state S (at rate γ). This same approach can be employed to account for movement of the population between any two states.

For this investigation, two routes of transmission are considered: primary transmission in which the exposure vehicle is background exposure and/or recreational contact in Newport Bay, and secondary transmission, which includes person-to-person transmission.

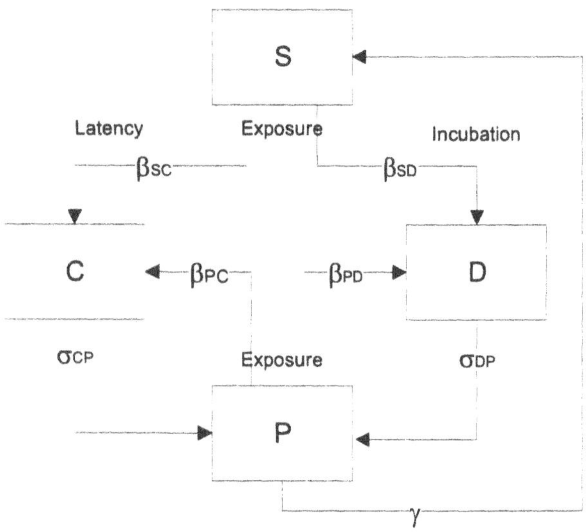


***State Variables***

S :Susceptible to Infection, not infectious, not symptomatic
C: Infectious and not symptomatic
D: Infectious and symptomatic
P: Protected from infection, not infectious, not symptomatic

***Rate Parameters***

$B_{sc}$: Fraction of individuals in state S that move to state C per hour
$B_{sd}$: Fraction of individuals in state S that move to state D per hour
$B_{pc}$: Fraction of individuals in state P that move to state C per hour
$B_{pd}$: Fraction of individuals in state S that move to state D per hour
$\sigma_{cp}$: Fraction of individuals in state C that move to state P per hour
$\sigma_{dp}$: Fraction of individuals in state D that move to state P per hour
γ:Fraction of individuals in state P that move to state S per hour

**Figure 1-2 Disease Transmission Model Conceptual Diagram**

Mathematically, the movement of the population between states may be modeled as a series of ordinary differential equations. Using this approach and assuming that each of the transmission processes described above are independent (Hethcote 1976), the change in the fraction of the population in any state from one time period to the next is modeled as a first order differential equation. The relative change in state S from one time period to the next due to primary infection is:

$$dS_1/dt = -\beta_{SC1}\, S - \beta_{SD1}\, S + \gamma\, P \qquad (1.2.6\text{-}1)$$

Similarly, the relative change in state S from one time period to the next due to person-to-person infections is directly related to the number of individuals who are in States S, C and D during that time period:

$$dS_2/dt = -(\beta_{SC2} + \beta_{SD2})\ S\ (D+C) \qquad (1.2.6\text{-}2)$$

Since primary and secondary transmissions are assumed to be independent processes (Hethcote 1976), the overall change in the number of susceptible individuals from one time period to the next is equal to $dS_1/dt + dS_2/dt$.

### 1.2.7 Characterization of Risk

Output from the disease transmission model includes the final number of people in each of the states as well as the average point prevalence for the simulation. Average point prevalence is defined as the proportion of the population (per 100,000) that is symptomatic (in state D) during each one-hour time step of the simulation averaged over the whole simulation period. Average point prevalence incorporates both the number of cases and the duration of disease, resulting in a measure of disease intensity. Average point prevalence can be compared with incidence (the number of new cases during a specified time period) by the following approximation (Eisenberg et al. 1996):

$$\text{Average Point Prevalence} \approx \text{Incidence} \bullet \text{Duration of disease.} \qquad (1.2.7\text{-}1)$$

Levels of background rotavirus disease reported in the literature were used as the feasible range for endemic disease from exposure to microorganisms in the community without the additional exposure from recreational activities in Newport Bay. A series of Monte Carlo simulation sets, composed of 1000 simulations each, were run employing the disease transmission model with variable levels of background exposure until a background exposure level was found that resulted in an average point prevalence consistent with median endemic levels of disease reported in the literature (Koopman et al. 1989; Mead et al. 1999). The identified background level of exposure was used in all subsequent simulations to ensure that the proportion of the population in each epidemiological state in the disease transmission model was an accurate estimation of the conditions in the community prior to additional exposure through recreational activity.

Exposure to enteric viruses in the Newport Bay watershed was simulated including background exposure and exposure through recreational activities. For each simulation, a set of values was selected randomly for each of the variables (Table 1-1), and the prevalence of disease due to background exposure was checked to ensure feasibility. If the prevalence fell outside of the feasible range, the parameter set was discarded and another set was randomly selected. If background prevalence fell within the feasible range, disease prevalence was computed for background exposure plus exposure through recreational activities. This process was repeated until 1000 feasible simulations were completed. The model was implemented using Matlab 5.3 and Simulink 3.0 (Mathworks Inc. 1998).

**Table 1-1. Model Parameter Summary**

| Model Parameters | Parameter | Units | Range | Source |
|---|---|---|---|---|
| Dose Response Parameters | Alpha | Unitless | 0.125 - 0.5 | Ward et al (1986) |
| | Beta | Unitless | 0.21 - 0.84 | Ward et al (1986) |
| Probability of Symptomatic Response | Psym | Unitless | 0.1 - 0.45 | Tufvesson et al (1977), Wenman et al. (1979), Haffejee (1995) |
| Previous Exposure Factor | $\varepsilon$ | Unitless | 0.1 - 0.9 | BPJ |
| Reciprocal of Incubation | taul | $day^{-1}$ | 0.33 -1.0 | Shepard et al. (1975), Flewett and Woode (1978) |
| Reciprocal of Latency | tauL | $day^{-1}$ | 0.143 - .333 | Gomez-Barreto et al. (1986), Ward et al. (1986) |
| Rate Diseased move to Post infection State | $\sigma$dp | $day^{-1}$ | 0.09 - 0.5 | Gomez-Barreto et al. (1986), Gurwith et al. (1981) |
| Rate Carriers move to Post infection State | $\sigma$cp | $day^{-1}$ | 0.05 - 0.125 | Ward et al (1986), Ansari et al. (1991) |
| Rate of susceptible re-establishment | $\gamma$ | $day^{-1}$ | 0.0009 - 0.002 | Bernstein et al. (1991), Ward and Bernstein (1994) |
| **Rate Parameters Dependant on Model Parameters** | | | | |
| $\beta_{SC}$ = (Pdose + Pcontact ) * (1-Psym) * tauL | | | | |
| $\beta_{SD}$ = (Pdose + Pcontact ) * (Psym) * taul | | | | |
| $\beta_{PC}$ = (Pdose + Pcontact ) * $\varepsilon$ * (1-Psym) * tauL | | | | |
| $\beta_{SD}$ = (Pdose + Pcontact ) * $\varepsilon$ * (Psym) * taul | | | | |
| **Intermediate Variables Used to Compute Rate Variables** | | | | |
| Psym = 1 - (1+ dose / Beta) ^ -Alpha | | | | Regli (1991) |
| Pcontact = 1.38 * $\sigma$ / N | | | | Hethcote (1976) |
| N = Population size | | | 1,200,000 | Site specific data |

# CHAPTER 2.0

# INTRODUCTION TO WATER QUALITY MODELING

Water quality modeling was carried out using RMA's finite element models of Newport Bay for simulation of fecal coliform and coliphage in support of the fecal coliform TMDL for the Newport Bay watershed. The water quality model provides fecal coliform and coliphage concentrations to the disease transmission model in time series format.

## 2.1 Background

The existing model of Upper Newport Bay was initially configured in 1997 and 1998 for hydrodynamics, sediment transport, and salinity simulation in support of the U.S. Army Corp of Engineers Feasibility Study (Resource Management Associates, Inc., 1997, 1998a, 1998b). The model was then extended in 1999 and 2000 for simulation of basic water quality parameters such as temperature, dissolved oxygen, nutrients, phytoplankton and macrophytes – with support of the California State Water Resources Control Board (Resource Management Associates, Inc., 2000).

The RMA model uses a finite element numerical model that permits explicit representation of the complex shape of Newport Bay and is well suited to simulation of shallow estuaries where large areas of mudflats are exposed at low tides. To date the focus of modeling efforts has been on the Upper Bay; however, the model does include the lower Bay to the ocean boundary. The RMA models support 1, 2, and 3-dimensional elements in a single computational network, but because the emphasis has been on the shallow Upper Bay, simulations have generally used a 2-dimensional depth averaged approximation. Calibration of the salinity model has shown that the 2-D depth averaged approximation is reasonable in the Upper Bay during much of the year. During and immediately following storm events, however, stratification is important in the Upper Bay. (Note that very little body contact recreation occurs in Newport Bay during the winter season when storms occur almost exclusively (EOA Inc. et al. 2001).

## 2.2 Objectives

The water quality modeling component of the health risk assessment was necessary for estimating pathogen dose as part of the exposure assessment. The goal of this water quality modeling effort was to extend the existing numerical modeling capability for Newport Bay to include simulation of fecal coliform and coliphage in support of the fecal coliform TMDL for the Newport Bay watershed. The numerical model provides temporally and spatially varying concentrations of fecal coliform and coliphage for use by the disease transmission model. Alternative control strategies that affect pathogen loading were also evaluated with regard to reduction of risk. Control strategies were investigated that were thought to have potential to decrease viral loads to the Bay. Specific objectives of the water quality modeling include the following:

- Configure and calibrate a fecal coliform and coliphage transport model based on the existing RMA model of Upper Newport Bay;
- Perform uncertainty analysis on model results;
- Develop a numeric linkage between the RMA model and the disease transmission model;
- Perform simulations for a set of alternative control strategies; and
- Provide the resulting coliphage concentration results to the disease transmission model.

## 2.3 Model Configuration

The RMA finite element model of Newport Bay extends from the tidal boundary at the entrance to the Lower Bay to San Diego Creek just upstream from Jamboree Bridge (Figure 2-1). Tributary inflows were specified at San Diego Creek, Santa Ana Delhi Channel, Big Canyon Wash, Back Bay Drain (calibration periods only), Fashion Island Drain, and Arches Drain. The Bay was represented using a two-dimensional depth averaged approximation, with short one-dimensional cross-sectionally averaged segments at the tidal boundary, San Diego Creek, and Santa Ana Delhi Channel.

The finite element mesh includes the area of the Bay up to approximately +1m (3.3ft) mean sea level (MSL). Because the bed elevations in the Bay change over time, several bathymetric data sets have been developed for the model. Bathymetric data representing current conditions, including the recent Unit III dredging, were used for this study. This data was based on the 1997 comprehensive survey performed by the County of Orange and the Unit III dredging design plans (County of Orange Public Facilities Resources Department 1997). A contour plot of the model bathymetry is shown in Figure 2-2. Model mixing coefficients were calibrated during a salinity study in Upper Newport Bay to best match observed salinity gradients in the Bay. These mixing coefficients were used for this study.

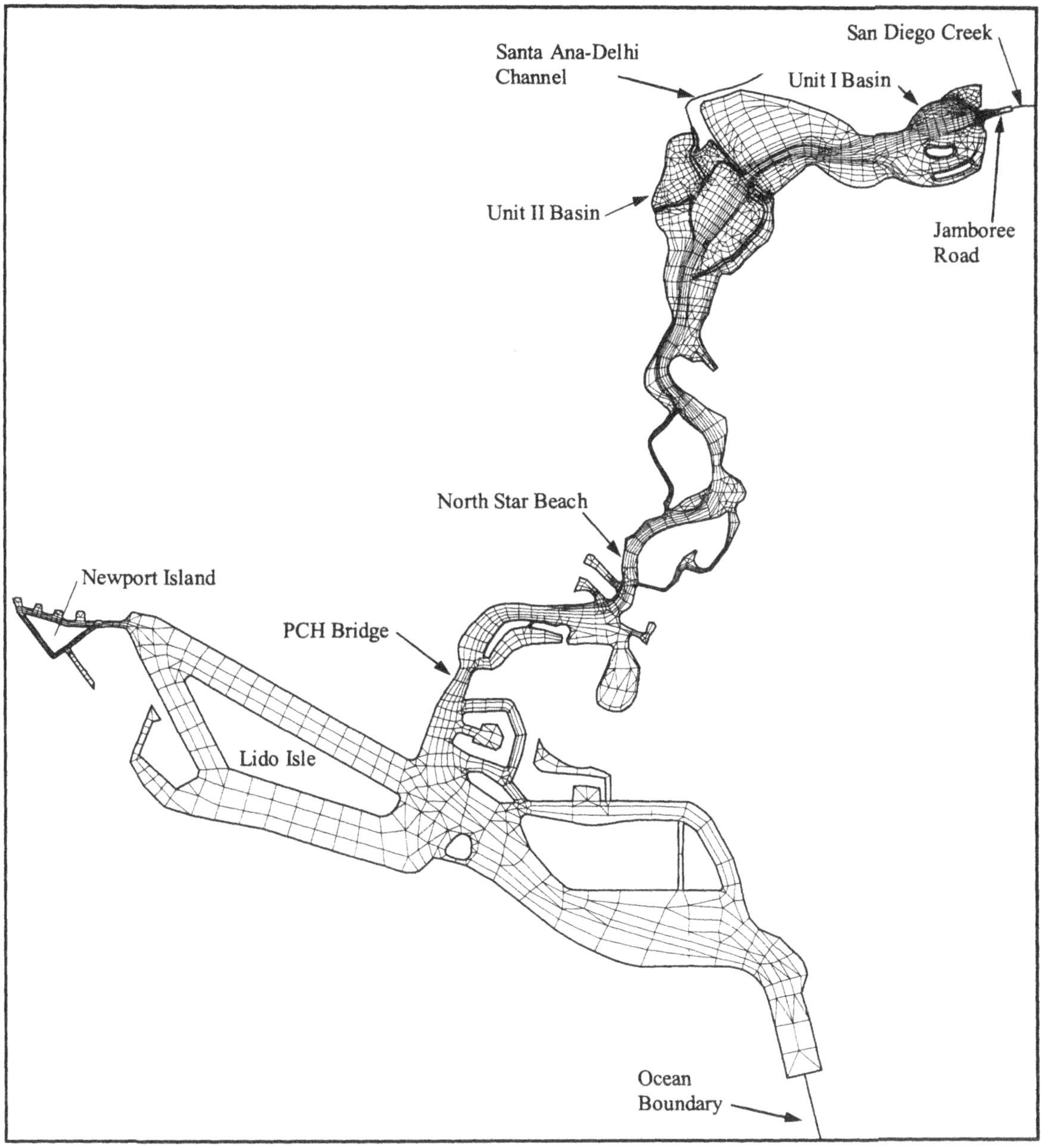


**Figure 2-1. Finite element configuration for Newport Bay, 1997 bathymetric conditions with Unit III basin constructed.**

Figure 2-2. Newport Bay bathymetric contours with Unit III basin constructed

CHAPTER 3.0

# WATER QUALITY MODELING: FECAL COLIFORM SIMULATIONS

Coliphage simulations are used directly in the health risk assessment, however, it was important to model fecal coliform to develop and test the loading strategy used in applying the RMA model of Newport Bay for pathogen transport because of the availability of coliform monitoring data. Only a very limited amount of coliphage data are available from the recent monitoring that was performed as part of the health risk assessment investigation. In contrast, fecal coliform monitoring has been part of the regular monitoring regime in the Bay for many years. Consistent with the health risk assessment methodology, dry season conditions were modeled because summer is the high recreation season and thus the period of highest exposure (EOA Inc. et al. 2001).

## 3.1 Boundary Conditions

Boundary conditions for the simulations include inflow, temperature and fecal coliform concentration at San Diego Creek, Santa Ana Delhi and other major drains, and temperature and tidal elevation at the ocean boundary (coliform concentration assumed zero at the ocean boundary). The location of the key areas is shown in Figure 3-1.

The Newport Dunes area is known to be the location most heavily used for body contact recreation in the Newport Bay Watershed (EOA Inc. et al. 2001). Bather fecal coliform loads were applied at perimeter elements in the Newport Dunes in the swimming area.

Boat discharge loads were applied at the marina locations as shown in Figure 3-2. Fecal coliform loads were also applied at perimeter elements throughout the Lower Bay (as shown in

Figure 3-1) to represent the mass loading from the small, unmonitored storm drains. The locations and mass loads were estimated based on location and size of the storm drains, and were fine-tuned to calibrate the model. A distributed load was applied to represent other unknown non-point sources.

Two coliform loss parameters were applied: die-off in darkness, and light sensitive die-off. Die-off rates are temperature dependent. The die-off rate in darkness ranges from 0.02 $hr^{-1}$ to 0.038 $hr^{-1}$, with temperature dependence. The depth-averaged light sensitive die-off at peak sunlight ranges from approximately 0.16 $hr^{-1}$ to 2.4 $hr^{-1}$, varying with depth and temperature. Literature values indicate a die-off rate in darkness of 0 – 0.1 $hr^{-1}$ (Hydroscience 1977) and a die-off rate in peak sunlight of 1.5 – 4.6 $hr^{-1}$ (Fujioka et al. 1981).

### 3.1.1 Flow

Average dry season flows were used for each of the creeks and drains. Boundary flows are shown in Table 3.1. Creek flow data were collected by the OCHCA and drain flow data were collected by Irvine Ranch Water District (IRWD).

### 3.1.2 Tide

Predicted tidal elevations for August 5 – September 5, 1998 were applied at the tidal boundary and are shown in Figure 3-3.

### 3.1.3 Fecal Coliform Data

Fecal coliform data were supplied for this investigation by OCHCA (refer to section 1.2.3). Sampling sites are shown in Figure 3.4. Boundary fecal coliform concentrations were set constant to the median of all available dry season fecal coliform data at each location. Data at the boundaries span ranges as wide as 4 to 5 orders of magnitude. Median values were chosen because averages of such a broad range of concentrations would be less meaningful. Boundary concentrations are summarized in Table 3-2.

Bather and boat loads were estimated for the health risk assessment and are documented in detail in the final report for that investigation (EOA, Inc., May 2001). Based on site-specific data collected for the health risk assessment, different loadings were used for weekdays and weekends. Bather loads vary by the type of day (weekday, weekend, or holiday) and by hour. The distributed load was constant throughout the Bay and was adjusted to bring the model to calibration. The distributed load and the ranges of bather and boat loads are presented in Table 3-1.

## 3.2 Calibration Results

The goal of the fecal coliform calibration was to match computed coliform concentrations with observed concentrations throughout the Bay as closely as possible.

**Table 3-1. Boundary condition flows and fecal coliform concentrations.**

| Source | Flow ($m^3$/sec) | Fecal Coliform |
|---|---|---|
| San Diego Creek | 0.4 | 1300 coliform/100mL |
| Santa Ana Delhi | 0.07 | 1300 coliform/100mL |
| Big Canyon Wash | 0.00997 | 400 coliform/100mL |
| Back Bay Drive | 0.00177 | 300 coliform/100mL |
| Fashion Island | 0.00157 | 3500 coliform/100mL |
| Arches Drain | 0.00656 | 3500 coliform/100mL |
| Bather Load (Dunes) | | 1.6E+7 – 1.1E+8 coliform/hr |
| Boat Load | □ | 9.88E+10 – 3.95E+11 coliform/day |
| Distributed Load | □ | coliforms/$m^2$□sec |

Water quality monitoring carried out by OCHCA included weekly fecal coliform observations for a period of more than 10 years. During that time period, monitoring occurred between the hours of 7 AM and 3 PM. The OCHCA database included as many as 400 dry weather observations at each sampling site. At most of the sampling locations, the fecal coliform concentrations have a range of three to four orders of magnitude. With data available only at weekly intervals, calibration for any exact time was not attempted. Instead, median dry season concentrations were used to set constant boundary conditions, and the model results were compared with median dry season fecal coliform concentrations within the Bay.

For each monitored site, computed fecal coliform concentration time series were generated along with the minimum, maximum, average and median dry season observed values. Representative results are shown in Figure 3-5 for the Bayshore Beach site. A complete set of time series plots for all sites is presented in Appendix A. It is important to note that virtually all of the data for each sampling station has a detection limit of 20 MPN/100mL. The three lowest observations recorded were 0, 20, and 40 MPN/100mL. Thus values of 0 MPN/100mL can be interpreted as being somewhere in the range of approximately 0 to 20 coliforms/100mL, and a value of 20 MPN/100mL can be interpreted as being in the range of approximately 20 to 40 MPN/100mL.

With OCHCA data samples taken only between the hours of 7 AM and 3 PM, it was not possible to determine if computed concentrations during the nighttime hours were consistent with in-Bay concentrations. The model computes large changes in coliform concentrations between pre-dawn and peak sunlight hours; thus it is inappropriate to compare computed results over 24 hours with observations from only daylight hours. For a more reasonable comparison, median and average values have been computed at each site for the model results between 7 AM and 3 PM only. These values are compared with the median observed values in Table 3-2. Comparison of computed and observed values for the daytime hours only is reasonable for this study because site-specific data indicate that the vast majority of body contact recreation occurs between the hours of 8 AM and 6 PM (EOA, Inc., 2001).

At most locations, the median and average computed values were in good agreement with the observed data. As shown in Table 3-2, at most locations the median computed values were within the range indicated by the observed data (e.g., acceptable range for an observed value of 20 MPN/100mL would be 20 to 40 MPN/100mL). The few locations that were not within the indicated range were within approximately 10% to 20% of the range, except at the Dunes, where the median computed value (15 MPN/100mL at Dunes East) was much lower than observed (130 MPN/100mL at Dunes East). At the Dunes East station, the computed values were lower than observed. Loads entering the Dunes area directly were bather load, Back Bay Drain, and the distributed load. Boat loading was applied just outside the Dunes area. These loads were not sufficient to explain the observed concentrations at the Dunes East site. There may be other non-point sources contributing to the Dunes that have not been included in the model, or the model may be over-predicting die-off rates. The die-off rates used in the model are consistent with literature values and produce good results throughout most of the Bay, however data were not available specifically to confirm that these rates are appropriate for Newport Bay. Another possible explanation for the noted discrepancy at the Dunes East site is under-representation in the model of the historic intermittent flow of the Back Bay Drain. Evidence suggests that the drain is a potential source of fecal coliform contamination. The Back Bay Drain was diverted to the sanitary sewer in 1999, and the concentrations of fecal coliform at the Dunes East site under the current conditions are in good agreement with model results.

**Table 3-2 . Observed and computed fecal coliform concentrations at sampling sites.**

| Site | Median Observed Fecal Coliform (MPN/100mL) | Median Computed Fecal Coliform (MPN/100mL) (7AM-3PM) | Average Computed Fecal Coliform (MPN/100mL) (7AM-3PM) |
|---|---|---|---|
| Ski Zone | 20 | 33 | 72 |
| Vaughn's Launch | 20 | 22 | 36 |
| North Star Beach | 20 | 18 | 27 |
| De Anza | 20 | 16 | 21 |
| Bayshore Beach | 0 | 17 | 21 |
| Promontory Point | 0 | 17 | 19 |
| Dunes North | 40 | 15 | 25 |
| Dunes West | 40 | 15 | 24 |
| Dunes Middle | 20 | 14 | 25 |
| Dunes East | 130 | 15 | 25 |
| Ruby | 0 | 17 | 20 |
| Onyx | 0 | 17 | 20 |
| Grand Canal | 20 | 18 | 21 |
| Park | 0 | 13 | 16 |
| Abalone | 0 | 11 | 14 |
| Sapphire | 0 | 12 | 14 |
| Garnet | 0 | 14 | 16 |
| Harbor Patrol Beach | 0 | 9 | 11 |
| Rocky Point | 0 | 5 | 7 |
| N Street Beach | 0 | 5 | 7 |
| Alvarado | 0 | 14 | 17 |
| 10th Street | 0 | 17 | 19 |
| 15th Street | 0 | 16 | 18 |
| Lido Isle | 0 | 16 | 18 |
| Via Genoa | 0 | 17 | 18 |
| 19th Street | 0 | 16 | 18 |
| Rhine Channel | 20 | 38 | 46 |
| 33rd Street | 40 | 23 | 31 |
| 38th Street | 0 | 27 | 34 |
| 43rd Street | 20 | 30 | 38 |

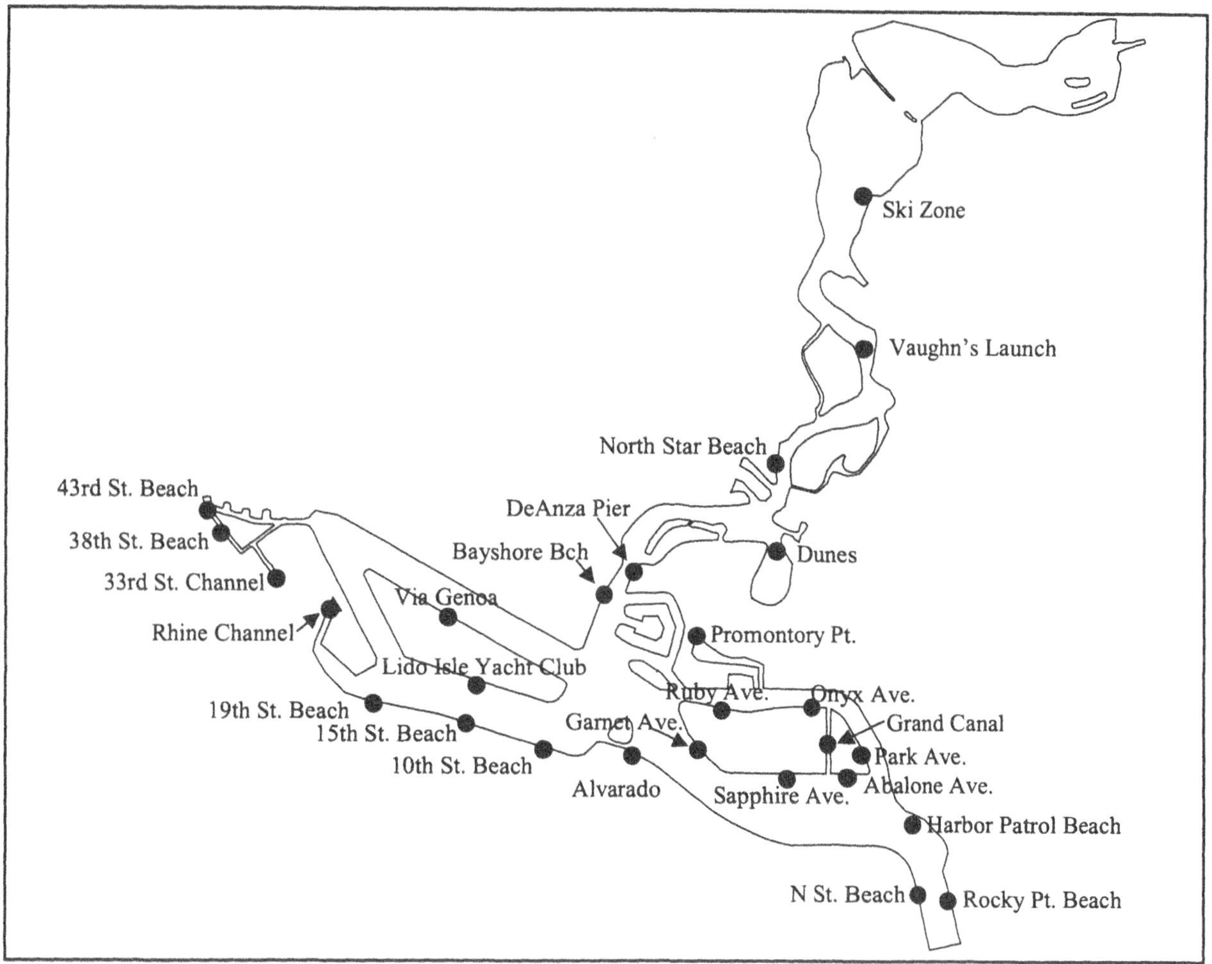


**Figure 3-1. Newport Bay key area map.**

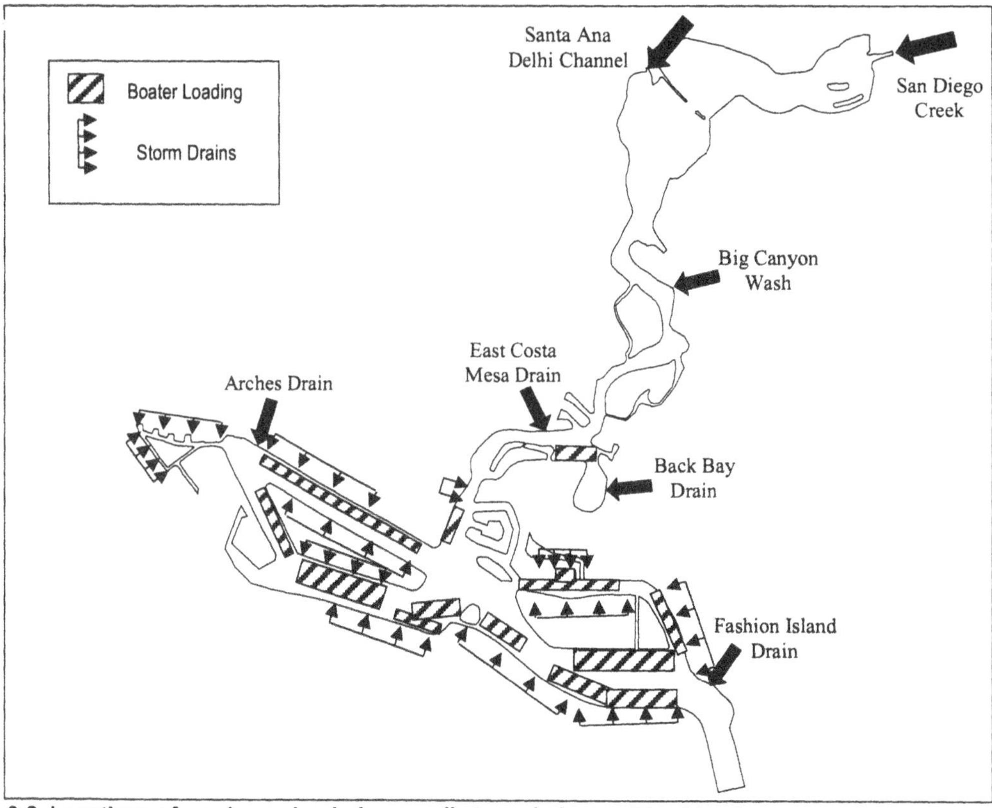


**Figure 3-2. Locations of creeks, major drains, small storm drains, and boat loads.**

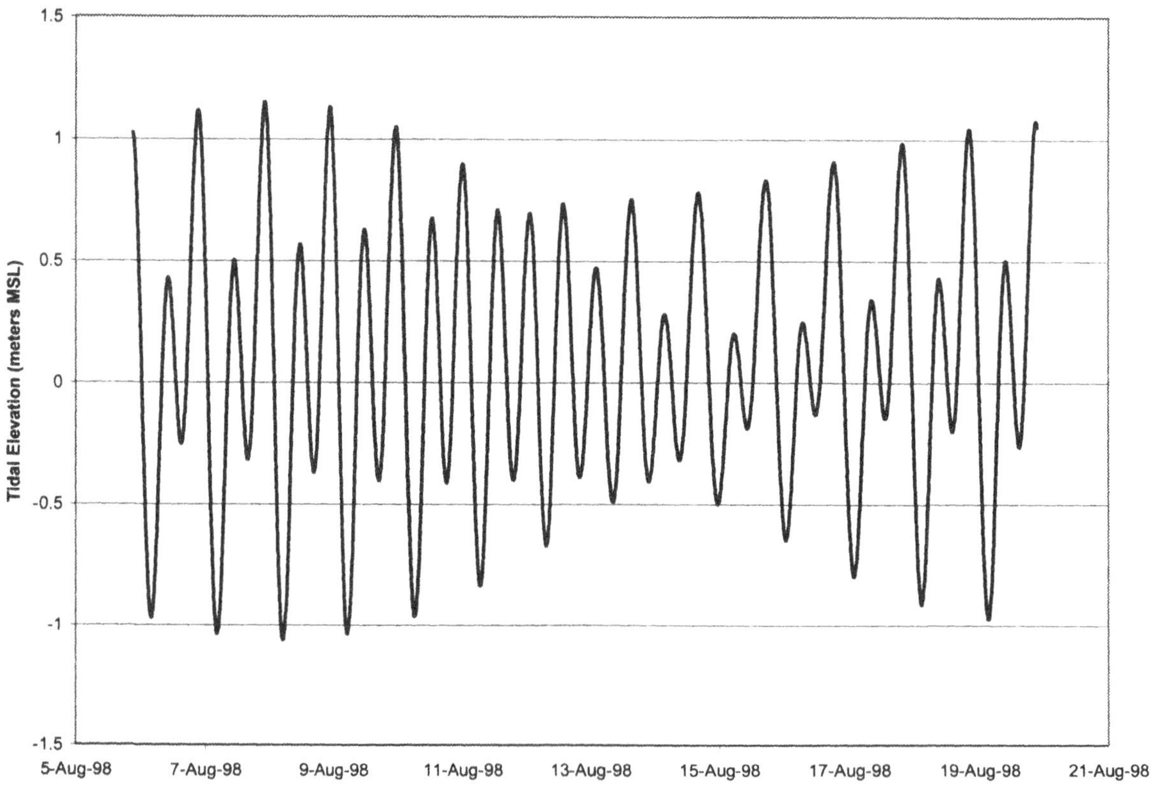


**Figure 3-3. Water surface elevations applied at the tidal boundary.**

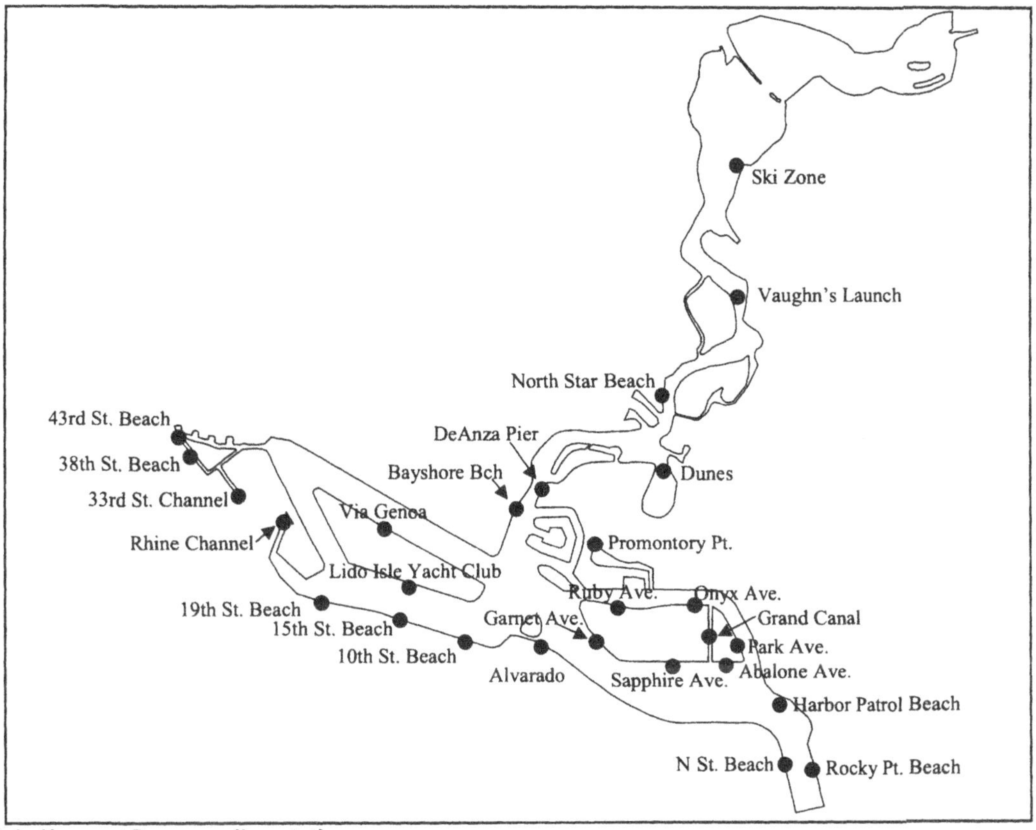


**Figure 3-4. Newport Bay sampling stations.**

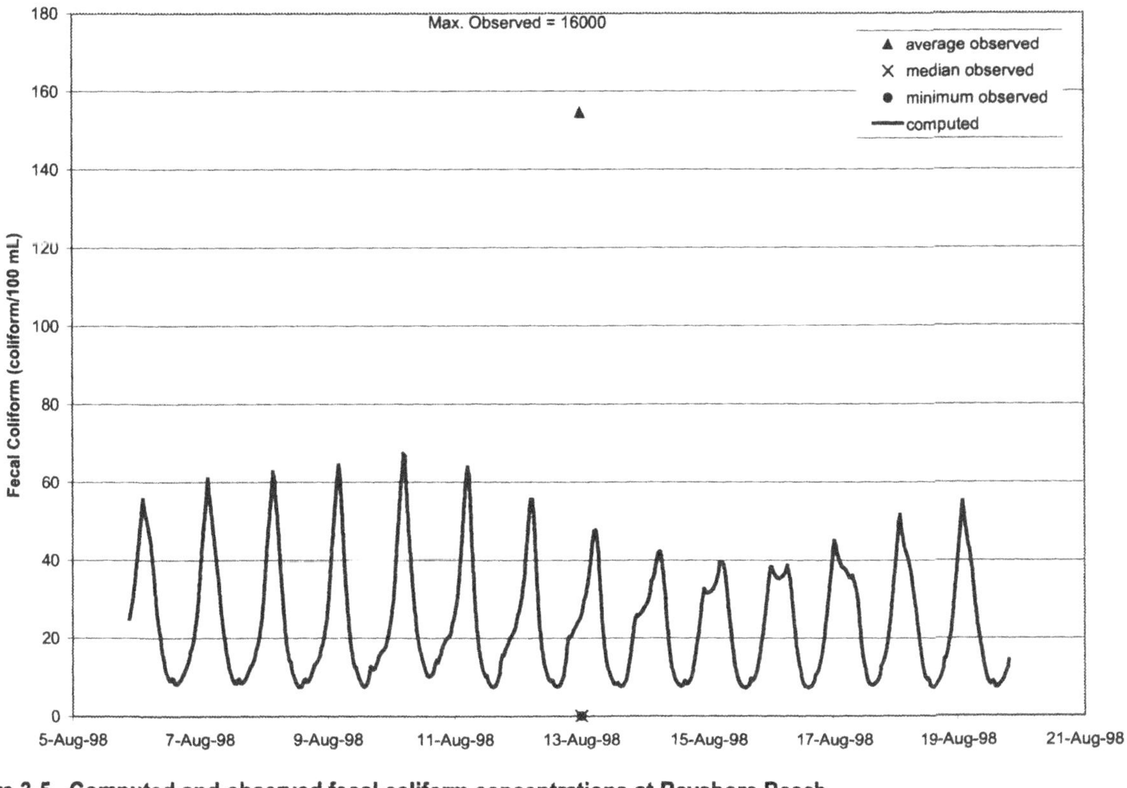


**Figure 3-5. Computed and observed fecal coliform concentrations at Bayshore Beach.**

CHAPTER 4.0

# WATER QUALITY MODELING: COLIPHAGE SIMULATIONS

The transport model for fecal coliform was reconfigured for coliphage using the same overall loading strategy. Loading and die-off rates were adjusted for coliphage. In the coliform simulations, the spatially distributed load and small storm drain loads were added to represent unmonitored fecal coliform, which are primarily of non-human origin. Because coliphage is assumed to be directly related to human sources and observed data were not sufficient to indicate whether additional coliphage loads may be entering the Bay from unmonitored sources, no additional coliphage loading was introduced through the spatially distributed load or small storm drains. As with the fecal coliform modeling efforts, only dry season conditions were modeled because summer is the high recreation season and thus the period of highest exposure (EOA, Inc., September 2000). The hydrodynamics used in the fecal coliform simulation were also used for the coliphage simulation.

## 4.1 Initial Conditions

Initial coliphage concentrations were set to 0.2 coliphage/100mL, although this initial concentration is of little significance because the model equilibrates after less than one day.

## 4.2 Boundary Conditions

Boundary conditions for the simulations include inflow, temperature and coliphage concentration at San Diego Creek, Santa Ana Delhi and the major drains, and temperature and tidal elevation at the ocean boundary (coliphage concentration assumed zero at the ocean boundary).

Bather virus loads were applied at perimeter elements in the Dunes where swimming occurs. Boat discharge virus loads were applied at the marina locations as shown in Figure 3-1.

Two coliphage loss parameters were applied: die-off in darkness, and light sensitive die-off. The die-off rates are temperature dependent. The die-off rate in darkness ranges from approximately 0.0024 to 0.0045 $hr^{-1}$, with temperature dependence. The depth-averaged light sensitive die-off at peak sunlight ranges from approximately 0.0028 to 0.041 $hr^{-1}$, varying with depth and temperature. Literature values indicate die-off rates ranging from 0.0028 $hr^{-1}$ to 0.096 $hr^{-1}$ (Fujioka et al. 1980; Hurst and Gerba 1980; Raphael et al. 1984).

Boundary coliphage concentrations were set to the maximum of all available dry season coliphage data at each location. Based on the ambient monitoring program carried out specifically for this investigation (refer to section 1.2.3), 1 to 12 dry season coliphage observations were available for each of the creeks and drains. The majority of these data were reported to be below detectable limits (<1 pfu/100mL). A detailed summary of the coliphage data collected for this investigation is provided in the health risk assessment final report (EOA Inc. et al. 2001). The maximum values were chosen for boundary concentrations to be health protective. If no data were above the detection limit for a particular inflow, the boundary was set at the detection limit. Boundary concentrations are shown in Table 4.1.

Bather and boat loads were estimated and are documented in detail in the health risk assessment final report (EOA Inc. et al. 2001). As described in the fecal coliform section, those locations vary by type of day and by the hour. The ranges of bather and boat loads are shown in Table 4-1.

**Table 4-1. Boundary condition flows and coliphage concentrations.**

| Source | Flow ($m^3$/sec) | Coliphage |
|---|---|---|
| San Diego Creek | 0.4 | 3 coliphage/100mL |
| Santa Ana Delhi | 0.07 | 6 coliphage/100mL |
| Big Canyon Wash | 0.00997 | 0.25 coliphage/100mL |
| Back Bay Drive | 0.00177 | 376 coliphage/100mL |
| Fashion Island | 0.00157 | 1 coliphage/100mL |
| Arches Drain | 0.00656 | 1 coliphage/100mL |
| Bather Load (Dunes) | □ | 6.1E+5 – 4.2E+6 coliphage/hr |
| Boat Load | □ | 3.09E+9 – 1.2E+10 coliphage/day |

## 4.3 Simulation Results

Loading for the coliphage simulations was configured using a conservative interpretation of the limited monitoring data, and in-Bay concentrations then were compared to observed data. Limited coliphage monitoring was carried out as part of the Ambient Monitoring Program summarized in Section 1.2.3. Two to five dry season coliphage observations were available at each station. Samples were taken between August 1999 and May 2000 (EOA Inc. et al. 2001). Almost all of the in-Bay data were reported to be below the detection limit (<1 pfu/100mL). Median values at all stations were below the detection limit. Based on the limited amount of quantifiable data from the site-specific monitoring program, rigorous calibration of the coliphage model was not possible. The model did produce in-bay concentrations consistent with the observed data where the predicted coliphage concentrations at the sampling sites were less than 1 coliphage/100 mL.

For the sampling sites shown in Figure 3-3, computed coliphage concentration time series were generated. Representative time series results for the Newport Dunes West sites are presented in Figure 4-1. The full set of coliphage time series plots for Newport Bay sites is presented in Appendix B. The first day of the results are not plotted on these graphs so that any impact of the initial condition can be disregarded. Computed values at all of these sites were less than 1 pfu/100 mL, which is a favorable comparison because the median observed coliphage concentrations at all stations were reported to be below the detection limit of 1 pfu/100 mL. A more rigorous calibration could be carried out in the future, should more quantifiable data become available.

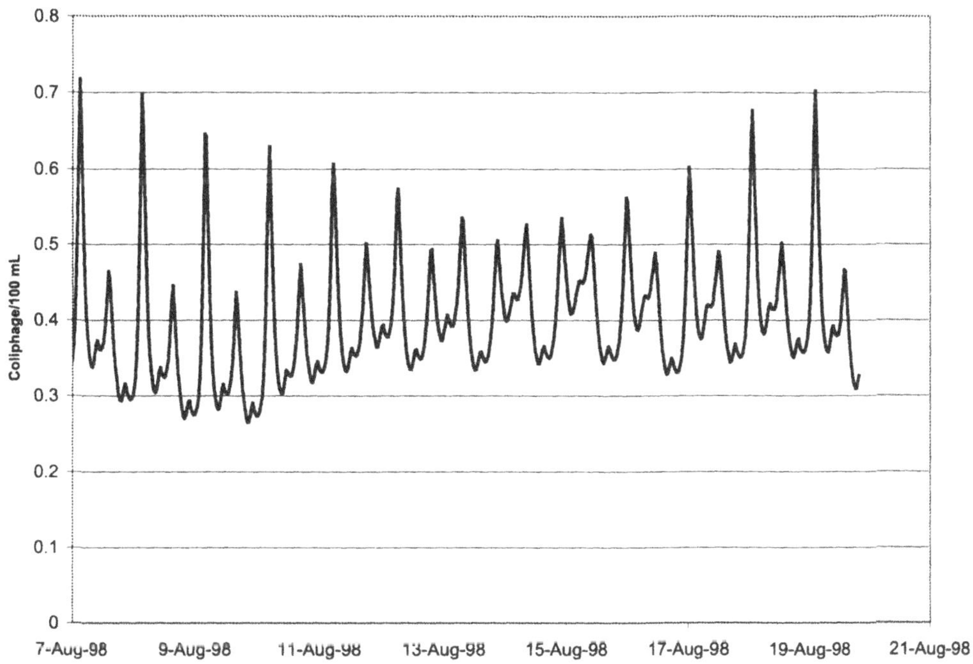


**Figure 4-1. Computed coliphage concentrations at Newport Dunes West.**

CHAPTER 5.0

# WATER QUALITY MODELING: ANALYSIS OF LOADING AND UNCERTAINTY

In previous applications, the RMA model of Newport Bay has been shown to successfully simulate transport of dissolved and suspended constituents when the sources (loading) of those constituents are reasonably well known (Resource Management Associates, Inc. 1997, 1998a, 1998b, 2000). The primary sources of uncertainty in the fecal coliform and coliphage modeling results are the estimates of loading and die-off rates. With respect to uncertainty analysis, the important issues to address are:

- The contribution of each source to the concentration at a particular site;
- The expected distribution of concentration at a particular site given the uncertainty in the estimate of each source; and
- The impact of the die-off rates on individual load contributions and expected distribution.

Because die-off is a first order decay process, linear superposition can be used to calculate the total concentration of fecal coliform and coliphage at any site. That is, thc total concentration at any site is equal to the sum of the concentrations resulting from each load individually. To examine the contribution of individual loads for a given die-off rate, separate simulations can be performed tracking the concentration response of individual loadings throughout the bay. The response is directly proportional to the individual loadings, so scaling the base response can generate the response for alternate individual loadings. A Monte Carlo methodology may be used to develop the distribution of the total concentration at a particular site by summing the response for each load scaled by the estimated load distribution. The impact of the die-off rate is bracketed by developing concentration responses to individual loads with different values for the die-off rates.

A post-processing routine was developed to determine the contribution from each load to total coliform and coliphage concentrations at 25 locations throughout the Bay, and to analyze the uncertainty of the model results based on the probability distributions of input data and uncertainty in die-off rates. The 25 locations for which results were analyzed are shown in Figure 3-3. All uncertainty analyses were performed using results between the hours of 8:00 and

18:00 because exposure via body contact recreation occurs almost exclusively between those hours (EOA, Inc. May 2001).

## 5.1 Loading Analysis Simulations

To develop the concentration response for each individual load, the coliform and coliphage simulations were performed for each load and three estimates of die-off rate. The die-off rate estimates include the following

- "Base" simulation = all parameters set the same as for the respective model calibration and historic simulations.
- "No light" simulation = no light dependent die-off.
- "Conservative" simulation = no die-off.

Each simulation was run for a 28-day period using a 14-day repeating spring-neap tidal cycle and summer low flow conditions.

Individual simulations were performed for each of 12 different loads/load types: (1) San Diego Creek, (2) Santa Ana Delhi Channel (3) Big Canyon Wash, (4) Arches Drain, (5) Fashion Island Drain, (6) Bather loading, (7) Vessel loading, (8) Distributed load, (9) Small storm drains in Newport Island area, (10) Small storm drains in Lido Isle area, (11) Small storm drains in Balboa Island area, and (12) Small storm drains in Linda Isle area. Flows and concentrations for each load type were set the same as for the calibration simulations (Back Bay Drain was not included as it has been diverted from the Bay for dry season flow).

For the "base" fecal coliform simulation, die-off rates were set near the low end of the literature value range. Even with these low rates, a large distributed load representing various non-point sources was required in the fecal coliform simulation to bring calibration concentrations up to fecal coliform concentrations observed in the Bay. As noted in Section 3.1, the distributed load was used as a calibration parameter to bring the fecal coliform results into the range of the observed data. For the "no light" and "conservative" simulations, computed in-Bay concentrations from every source were increased. To illuminate the relative importance of individual loads, the distributed load was reduced for the "no light" fecal coliform uncertainty analysis simulation and the distributed load was eliminated for the "conservative" simulation. Decreasing or eliminating the contributions from these loads kept the resulting in-Bay fecal coliform concentrations in the range of the observed data. This is necessary because if the die-off rate were lower, the tributary, drain, vessel, and bather loads would be responsible for a larger proportion of the total concentration of fecal coliform at any given site. This possibility has been investigated because it is uncertain whether die-off rates taken from literature values are applicable to Newport Bay.

## 5.2 Individual Load Contributions

Results from each set of individual load simulations were post-processed to determine the average concentration attributable to each load type at each of the 25 sites of interest. The results of this analysis are presented in Figures 5-1 and 5-2 for the coliform and coliphage simulations, respectively.

### 5.2.1 Coliform

For the "base" coliform simulation, the distributed load makes up the majority of the total concentration at each location in the Bay (Figure 5-1). There were no observed data available to indicate actual loads from the small storm drains or the distributed load, thus these loads were estimated and adjusted in the calibration process to bring the computed in-Bay coliform concentrations into agreement with observed data. If the coliform die-off rate used in the "base" simulation was too high, then the importance of the small storm drain and distributed loads are over estimated, compensating for the coliform from other loads, which would have been eliminated too rapidly. If lower die-off rates are used in the simulation, as is the case for the "no light" and "conservative" runs, then the other, documented, loads will make up a greater proportion of the total coliform concentration in the water column, and the small storm drain and distributed load required to bring the simulated coliform concentrations up to observed values will be less. For the "no light" simulation, the calibrated distributed load and the small Newport Island storm drain loads were scaled down so that total concentrations remained near the observed concentrations. In the "conservative" case, the summed concentration from the known loads was higher than observed values. The addition of the distributed load and small storm drain loads would increase the concentrations in the Bay even higher above observed values, so those loads were eliminated. The "conservative" simulation indicates that without the distributed load, the San Diego Creek load makes up the largest portion of the total concentration. Santa Ana Delhi Channel and vessel loads contribute significantly as well.

Maximum, average, and minimum individual load contributions for each of the coliform simulations are presented for the Newport Dunes in Table 5-1. This table enables comparisons among the individual load contributions for the three different simulation scenarios. Results at the Dunes are tabulated as an example; results vary by location throughout the Bay. Removal of the light dependent die-off shifts the load contribution at the Dunes from the distributed load toward the other loads, mainly the creeks, vessel load and bather load. Removing all die-off and simulating coliform as a conservative constituent eliminates any contribution from the distributed and small Newport Island drain loads and increases the creek loads by the largest amount.

### 5.2.3 Coliphage

Coliphage results are presented in Figure 5-2. Note that the small storm drains and distributed loads were not included in the coliphage model because no data were available for them and because in-Bay data were not sufficient to indicate whether there might be additional loads entering the Bay from unmonitored sources. The vessel loading contributes the most to the total coliphage concentration at each of the sites. This is because it is the largest load and because it occurs in the vicinity of the 25 sites that were analyzed. The San Diego Creek and Santa Ana Delhi loads are nearly as large and are each input to only one element, whereas the vessel load is spread out over several different areas. The creeks produce higher concentrations

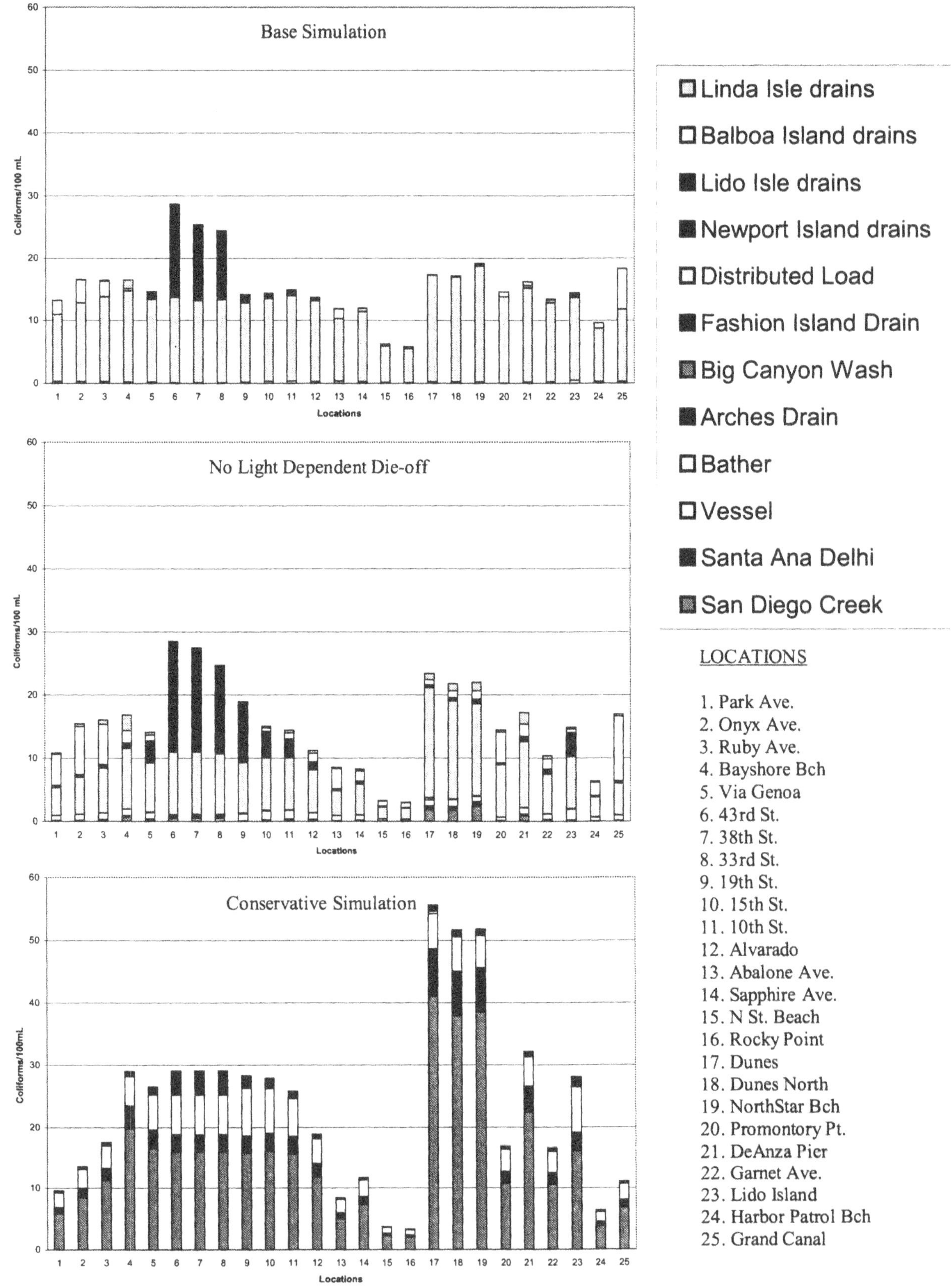


**Figure 5-1 Load contributions to total coliform concentrations at Newport Bay sites**

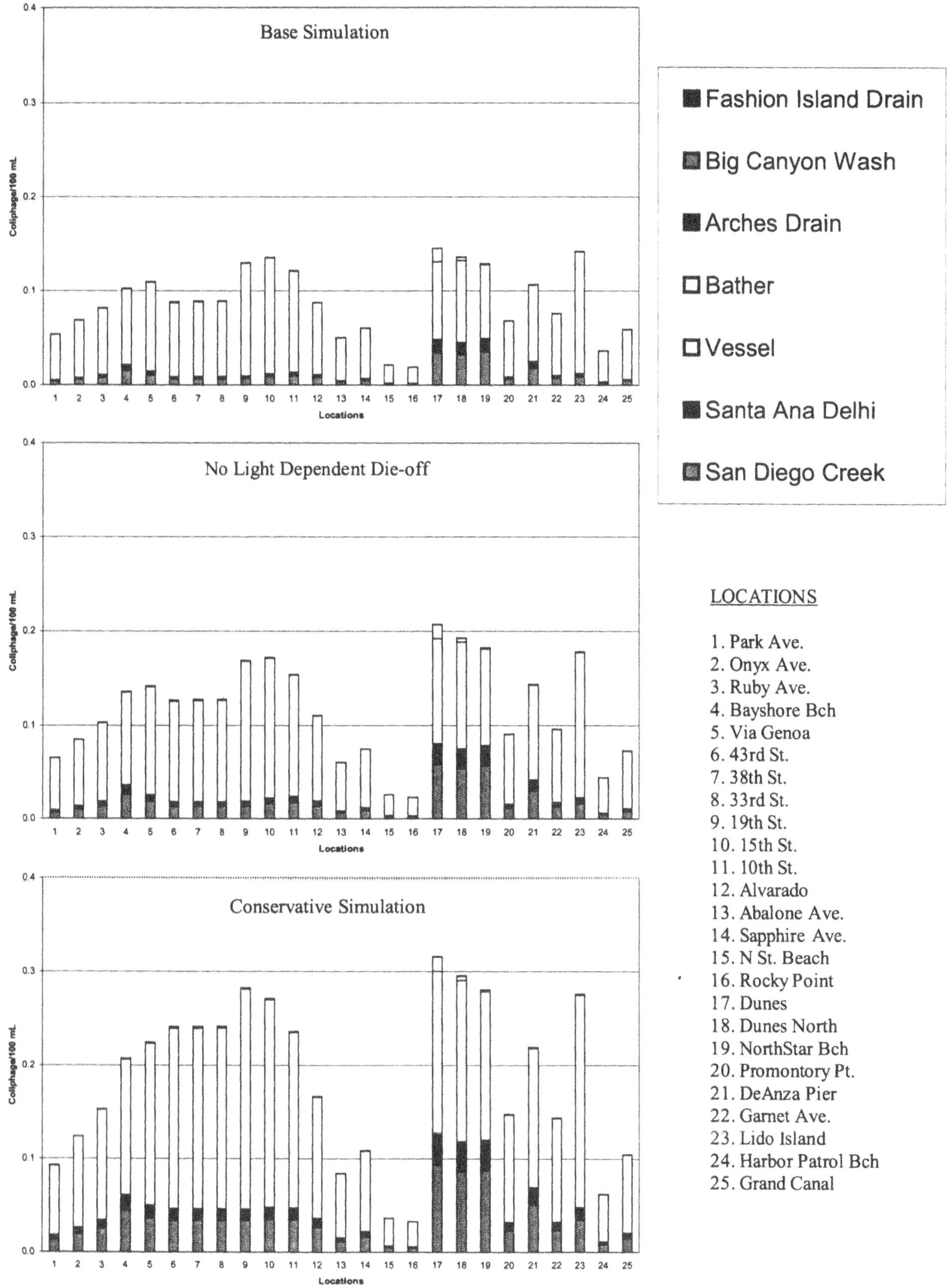


**Figure 5.2 Load contributions to coliphage concentrations at Newport Bay sites**

near the locations where they enter the Bay, however none of the 25 sites of interest are located in the areas where concentrations are elevated by the creek loads. The vessel loading could be less significant if the small storm drains were included; however, with the limited in-Bay concentration data and lack offlow or concentration data for the small storm drains, the possible impact of the small storm drain loadings cannot be assessed.

Maximum, average, and minimum individual load contributions for each of the coliphage simulations are presented in Table 5-2. This table enables comparisons among the individual load contributions for the three different simulation scenarios. Elimination of the light dependent die-off increases concentrations from individual loads at the Dunes by up to 1.7 times, with the largest increases seen in the San Diego Creek contribution. The conservative simulation increases concentrations from individual loads at the Dunes by up to 2.7 times, and again the largest increase was in the San Diego Creek contribution. These results vary for different sites throughout the Bay.

## 5.3 Uncertainty Analysis Methodology

A routine was developed to analyze the uncertainty of the model results based on the probability distributions of input data and uncertainty in die-off rates. A Monte Carlo simulation was performed for each of the six sets of coliform and coliphage analysis simulations, selecting values from each probability distribution for each load. The base analysis simulation results were scaled according to the selected distribution value and the concentrations from the individual loads were summed to produce the total concentrations at 25 locations throughout the Bay. Maximum, minimum and average values for each of these locations were tabulated.

For the uncertainty analysis it was necessary to represent each load type with a probability distribution. To develop distributions for the inflow fecal coliform concentrations, dry season data from the OCHCA datasets for San Diego Creek and Santa Ana Delhi Channel were used because these datasets contain the greatest number of observations, allowing for better definition of the probability distribution. Of the inflows input in the model, San Diego Creek and Santa Ana Delhi Channel were the only two for which fecal coliform data were available in the OCHCA dataset. These data were best represented by a log-normal probability distribution. Fecal coliform data were available from IRWD from data collected as part of the Health Risk Assessment ambient monitoring program (EOA Inc. et al. 2001), however the dry season data for the drains (Big Canyon Wash, Arches Drain, and Fashion Island) were not sufficient for development of reasonable probability distributions. Therefore, the San Diego Creek distribution was applied to the drains mentioned above, and to the small Lower Bay storm drains for which no data were available. The bather and vessel fecal coliform loads were represented with a log-uniform distribution ranging between the minimum and maximum literature values. The range of distributed load was examined by reducing or eliminating it for the no light dependent die-off and conservative simulations, respectively. The distributed load was held constant at the calibrated value for the base simulation uncertainty analysis, at the reduced value for the no light dependent die-off simulation uncertainty analysis, and eliminated for the conservative simulation uncertainty analysis.

Limited data were available for development of coliphage probability distributions as part of the Health Risk Assessment ambient monitoring program (EOA Inc. et al. 2001). One to twelve coliphage concentration data points were collected for each of the creeks and major drains. The majority of the data points were below detection limit (<1 coliphage/100mL). Uniform distributions were used for each of the creeks and major drains, ranging from 0 to the maximum observed value. If there were no values above detection limit, the maximum value was set at 1 coliphage/100mL. The distribution from Santa Ana Delhi Channel had the largest range (ranging from 0 to 6). The vessel coliphage loads were represented with a log-uniform distribution ranging between the minimum and maximum literature vales. Bather coliphage loads were represented with a log-normal probability distribution. The distributed load and the unmonitored storm drains were not included in the uncertainty analysis, nor were they included in the calibration or alternative scenarios.

Table 5-3 summarizes the coliform and coliphage base loads, and input data distribution statistics for each load type for the uncertainty analysis. The same coliform and coliphage input data distribution statistics were used for each of the different die-off rate scenarios.

## 5.4 Uncertainty Analysis Results

The final result from the Monte Carlo simulation was a probability distribution at each of the 25 sites. These distributions describe the probability and range of uncertainty in the model results due to variation in the input loads. The most probable, median, maximum, and minimum concentrations from the distributions for each site are tabulated. The probability distribution statistics were computed on log transformed data (i.e., the probability distribution was computed using the logs of the concentrations) which were then converted back to the base values for plotting and tabulation.

### 5.4.1Coliform

As an example illustration of the coliform uncertainty analysis results, each of the coliform Monte Carlo simulation probability distributions at the Dunes is shown in Figure 5-3. The varying shapes of the probability distributions shown in Figure 533 reflect the variation in the small storm drain loads and the distributed load. The distributed load and small Newport Island storm drain loads were scaled down for the simulation without light dependent die-off. The distributed load and all small storm drain loads were eliminated for the conservative simulation. Note that the conservative simulation is not plotted on the same probability scale as the upper two plots. The statistics for each of the 25 sites are summarized in Tables 5-4 through 5.6 for the three coliform simulations. The maximum expected values are 1.2 to 3.7 times the computed base simulation values. Maximum expected values are 3.1 to 23.5 times the computed no light dependent die-off simulation values, and 76 to 108 times the computed conservative simulation values. Elimination of light dependent die-off and subsequent reduction of the distributed and small Newport Island storm drain loads results in most probable concentrations ranging from 0.9 to 3.6 times the base simulation values. With no die-off and removal of the distributed load and all small storm drain loads, the most probable concentrations for the conservative simulation range from 1 to 8.5 times the base simulation values. The largest impact of the decreased die-off scenarios was seen at the Dunes.

### 5.4.2 Coliphage

The large range of uncertainty in the vessel loads (6 orders of magnitude) dominates the maximum range of the coliphage probability distribution. While the base simulation results were less than 1 coliphage/100mL (with the average vessel load as input), the maximum Monte Carlo simulation results were in the hundreds due to the high maximum range in vessel loads. Example plots of coliphage Monte Carlo simulation probability distributions at the Dunes are shown in Figure 5-4 for the three coliphage analysis simulations. The statistics for each of the 25 sites are summarized in Tables 5-7 through 5-9 for the three simulations. The maximum expected values are approximately 200 to 600 times larger than the computed maximum values for all simulations. Elimination of light dependent die-off increases the most probable concentration by 1.6 to 2 times. For the conservative simulation, the most probable concentrations are increased by about 2.2 to 4.7 times.

## 5.5 Future Extension of Uncertainty Analysis

The uncertainty analysis results presented here are based on estimates of the loading distributions. It is possible that in the future other interested parties may want to compute loads based on different estimates for the loading distributions. The post-processing routine and associated model data are applicable for other loadings and may be applied in a straightforward manner in the future should new data become available.

**Table 5-1 Coliform concentration contributions from each load type at the Newport Dunes**

| Load Type | Base Simulation | | | No light-dependent die-off | | | Conservative | | |
|---|---|---|---|---|---|---|---|---|---|
| | Max | Avg | Min | Max | Avg | Min | Max | Avg | Min |
| | Coliforms/100 mL | | | Coliforms/100 mL | | | Coliforms/100 mL | | |
| San Diego Creek | 0.320 | 0.0219 | 0.0000 | 3.35 | 1.85 | 0.784 | 56.5 | 41.0 | 25.1 |
| Santa Ana Delhi | 0.207 | 0.0163 | 0.0000 | 1.02 | 0.613 | 0.318 | 10.6 | 7.63 | 4.48 |
| Vessel Load | 0.776 | 0.0676 | 0.0011 | 1.76 | 0.923 | 0.468 | 7.29 | 5.57 | 3.74 |
| Bather Load | 0.351 | 0.127 | 0.0013 | 0.863 | 0.314 | 0.0210 | 0.992 | 0.392 | 0.0617 |
| Arches Drain | 0.0007 | 0.0001 | 0.0000 | 0.0176 | 0.0102 | 0.0042 | 0.726 | 0.621 | 0.537 |
| Big Canyon Wash | 0.0215 | 0.0026 | 0.0000 | 0.0577 | 0.0445 | 0.0337 | 0.399 | 0.313 | 0.232 |
| Fashion Island Drain | 0.0006 | 0.0001 | 0.0000 | 0.0058 | 0.0044 | 0.0029 | 0.0608 | 0.0522 | 0.0452 |
| Distributed Load | 84.0 | 17.0 | 5.4852 | 20.5 | 17.4 | 15.4 | 0.0000 | 0.0000 | 0.0000 |
| Newport Island Drains | 0.0014 | 0.0001 | 0.0000 | 0.0267 | 0.0147 | 0.0055 | 0.0000 | 0.0000 | 0.0000 |
| Lido Isle Drains | 0.0636 | 0.0070 | 0.0001 | 0.574 | 1.63 | 0.242 | 0.0000 | 0.0000 | 0.0000 |
| Balboa Island Drains | 0.161 | 0.0212 | 0.0002 | 1.05 | 3.36 | 0.603 | 0.0000 | 0.0000 | 0.0000 |
| Linda Isle Drains | 0.405 | 0.0587 | 0.0003 | 1.12 | 3.84 | 0.758 | 0.0000 | 0.0000 | 0.0000 |

**Table 5-2 Coliphage concentration contributions from each load type at the Newport Dunes**

| Load Type | Base Simulation | | | No light-dependent die-off | | | Conservative | | |
|---|---|---|---|---|---|---|---|---|---|
| | Max | Avg | Min | Max | Avg | Min | Max | Avg | Min |
| | Coliphage/100 mL | | | Coliphage/100 mL | | | Coliphage/100 mL | | |
| San Diego Creek | 0.0495 | 0.0340 | 0.0202 | 0.0796 | 0.0578 | 0.0314 | 0.1302 | 0.0927 | 0.0482 |
| Santa Ana Delhi | 0.0199 | 0.0141 | 0.0091 | 0.0304 | 0.0224 | 0.0135 | 0.0487 | 0.0348 | 0.0183 |
| Vessel Load | 0.1163 | 0.0827 | 0.0478 | 0.1445 | 0.1117 | 0.0715 | 0.2102 | 0.1725 | 0.1163 |
| Bather Load | 0.0363 | 0.0141 | 0.0018 | 0.0375 | 0.0149 | 0.0020 | 0.0385 | 0.0156 | 0.0022 |
| Arches Drain | 0.0002 | 0.0001 | 0.0000 | 0.0003 | 0.0001 | 0.0001 | 0.0004 | 0.0002 | 0.0002 |
| Big Canyon Wash | 0.0001 | 0.0001 | 0.0001 | 0.0002 | 0.0001 | 0.0001 | 0.0002 | 0.0002 | 0.0001 |
| Fashion Island Drain | 0.0000 | 0.0000 | 0.0000 | 0.0001 | 0.0000 | 0.0000 | 0.0001 | 0.0000 | 0.0000 |

**Table 5-3 Coliform and coliphage probability distribution statistics and base calibration concentrations.**

| Load Type | Distribution Type | Base Value | Max. Data Range | Min. Data Range | Log Mean Data Value | Log Standard Deviation |
|---|---|---|---|---|---|---|
| **Coliform** | | MPN/100mL* | | | | |
| San Diego Creek | Log-normal | 1300 | 160000 | 0 | 3.12 | 0.88 |
| Santa Ana Delhi | Log-normal | 1300 | 160000 | 0 | 3.01 | 0.99 |
| Big Canyon Wash | Log-normal | 400 | 160000 | 0 | 3.12 | 0.88 |
| Arches Drain | Log-normal | 3500 | 160000 | 0 | 3.12 | 0.88 |
| Fashion Is. Drain | Log-normal | 3500 | 160000 | 0 | 3.12 | 0.88 |
| Bather | Log-uniform | 1.2E+6 | 2.27E+6 | 1.2E+5 | -- | -- |
| Vessel | Log-uniform | 9.88E+10 | 5E+11 | 6E+9 | -- | -- |
| Newport Is. drains | Log-normal | 4000 | 160000 | 0 | 3.12 | 0.88 |
| Lido Isle drains | Log-normal | 4000 | 160000 | 0 | 3.12 | 0.88 |
| Balboa Is. drains | Log-normal | 4000 | 160000 | 0 | 3.12 | 0.88 |
| Linda Isle drains | Log-normal | 4000 | 160000 | 0 | 3.12 | 0.88 |
| Distributed | Constant | 20 coliforms/ $m^2$□sec | 20 coliforms/ $m^2$□sec | 0 | -- | -- |
| **Coliphage** | | Coliphage/100mL* | | | | |
| San Diego Creek | Uniform | 3 | 3 | 0 | -- | -- |
| Santa Ana Delhi | Uniform | 6 | 6 | 0 | -- | -- |
| Big Canyon Wash | Uniform | 0.25 | 1 | 0 | -- | -- |
| Arches Drain | Uniform | 1 | 1 | 0 | -- | -- |
| Fashion Is. Drain | Uniform | 1 | 1 | 0 | -- | -- |
| Bather | Log-normal | 4.6E+4 | 6.3E+7 | 2.9E+1 | 4.65 | 0.97 |
| Vessel | Log-uniform | 3E+9 | 5E+12 | 3E+6 | -- | -- |

*Units do not apply to distributed load.

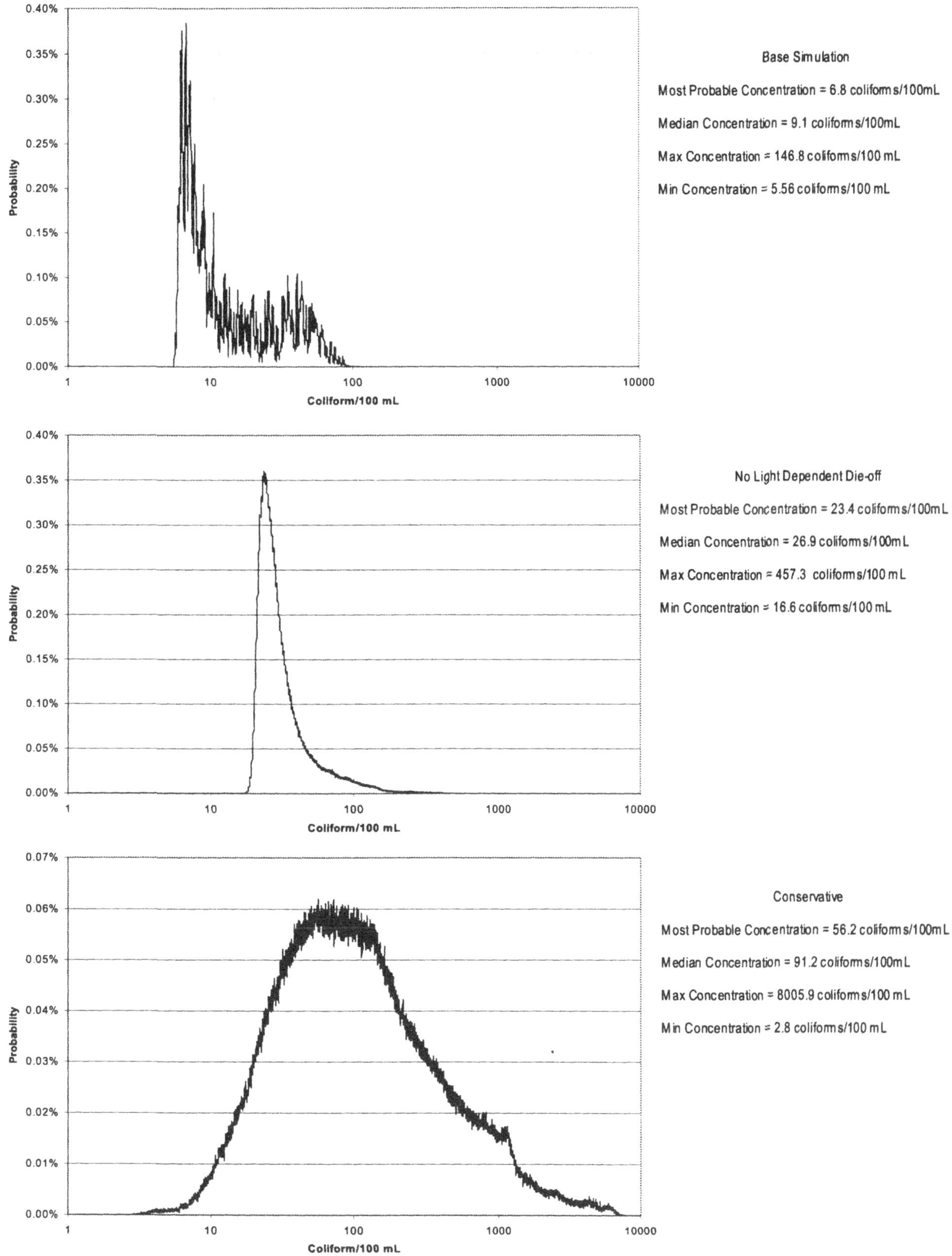


**Figure 5-3 Coliform probability distribution at the Dunes resulting from Monte Carlo simulations**

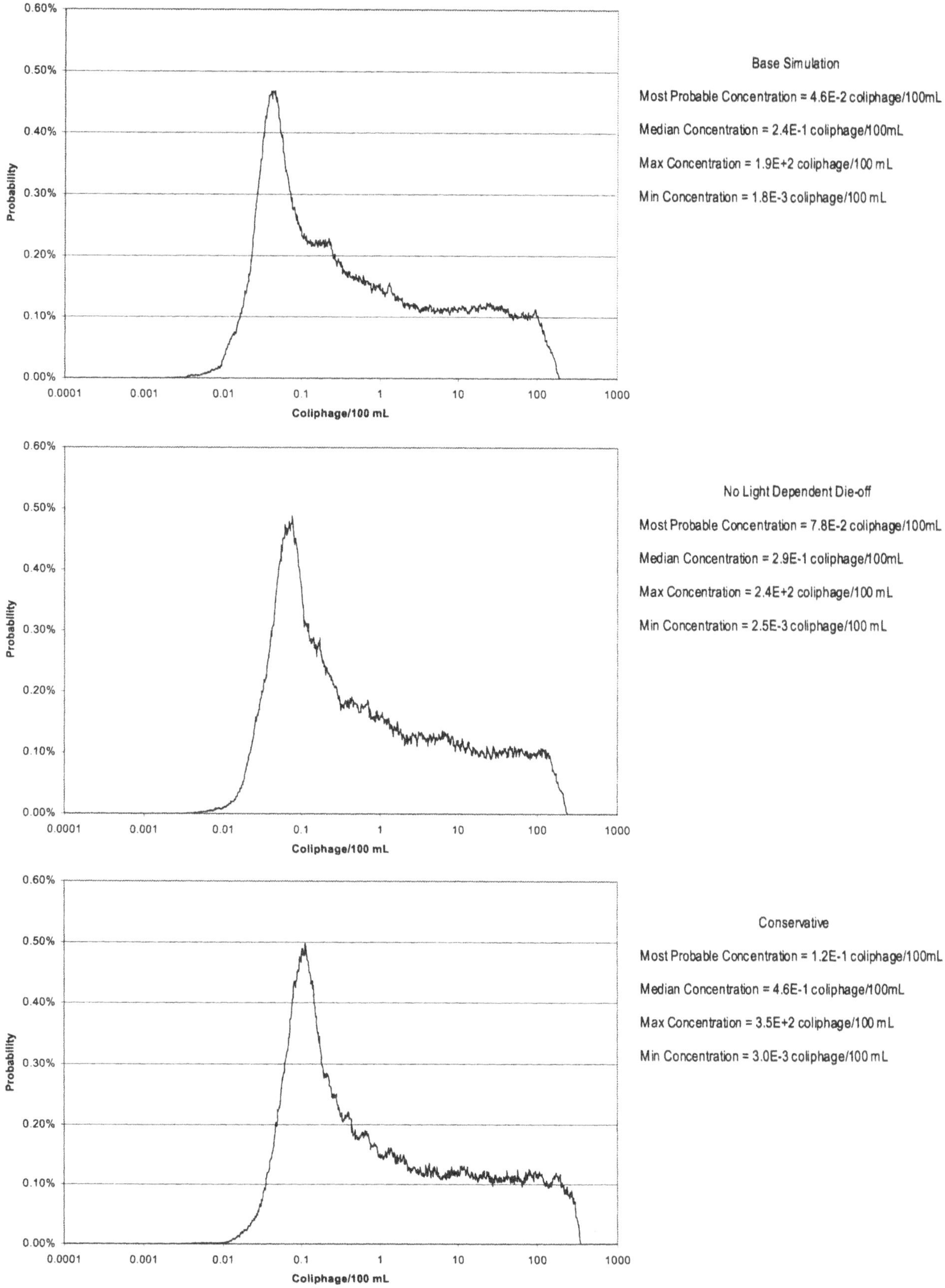


**Figure 5-4 Coliphage probability distribution at the Dunes resulting from Monte Carlo simulations**

**Table 5-4 Monte Carlo simulation results for base coliform simulation.**

| Site Name | Base Simulation Max | Avg | Min | Monte Carlo Most Prob. | Median | Max | Min |
|---|---|---|---|---|---|---|---|
| | Coliform/100 mL | | | Coliform/100 mL | | | |
| Park | 36.5 | 13.3 | 6.64 | 9.55 | 12.0 | 66.9 | 5.90 |
| Onyx | 41.2 | 16.6 | 8.89 | 11.0 | 15.1 | 79.1 | 7.27 |
| Ruby | 43.8 | 16.5 | 7.75 | 9.33 | 13.8 | 73.4 | 7.07 |
| Bayshore | 54.4 | 16.5 | 6.94 | 7.94 | 12.9 | 99.9 | 6.48 |
| ViaGenoa | 35.2 | 14.6 | 6.93 | 8.13 | 12.3 | 46.0 | 6.47 |
| 43rdStreet | 86.4 | 28.7 | 12.4 | 16.6 | 24.5 | 289 | 6.59 |
| 38th Street | 82.6 | 25.4 | 10.4 | 14.1 | 20.4 | 269 | 6.24 |
| 33rd Street | 76.3 | 24.4 | 10.4 | 13.8 | 20.4 | 236 | 6.26 |
| 19th St | 37.7 | 14.2 | 5.91 | 6.61 | 10.7 | 63.9 | 5.80 |
| 15th St | 38.1 | 14.4 | 6.29 | 7.08 | 11.0 | 47.9 | 6.02 |
| 10th St | 40.3 | 14.9 | 6.71 | 7.41 | 11.7 | 48.0 | 6.39 |
| Alvarado | 37.3 | 13.7 | 6.75 | 7.59 | 11.0 | 44.4 | 6.72 |
| Abalone | 34.2 | 11.9 | 5.77 | 9.12 | 10.7 | 52.8 | 5.28 |
| Sapphire | 33.0 | 12.0 | 6.13 | 7.76 | 10.0 | 43.7 | 6.07 |
| N Street Beach | 26.1 | 6.23 | 1.44 | 7.24 | 5.25 | 34.8 | 1.44 |
| Rocky Pt | 25.5 | 5.81 | 1.21 | 7.24 | 4.68 | 34.0 | 1.21 |
| Dunes | 86.3 | 17.3 | 5.49 | 6.76 | 9.12 | 147 | 5.56 |
| Dunes N | 78.6 | 17.1 | 5.63 | 6.17 | 9.33 | 168 | 5.63 |
| North Star Beach | 81.7 | 19.1 | 6.16 | 7.24 | 11.2 | 304 | 6.20 |
| Promontory Point | 37.0 | 14.6 | 6.11 | 6.76 | 11.0 | 48.7 | 6.05 |
| DeAnza Pier | 58.0 | 16.2 | 6.55 | 7.41 | 11.5 | 106 | 6.35 |
| Garnet Ave | 35.9 | 13.4 | 6.76 | 7.41 | 10.7 | 45.6 | 6.72 |
| Lids Island | 38.5 | 14.5 | 6.33 | 7.08 | 11.2 | 45.8 | 6.02 |
| Harbor Patrol Beach | 31.5 | 9.65 | 3.77 | 8.51 | 8.71 | 45.8 | 3.63 |
| Grand Canal | 46.1 | 18.3 | 9.98 | 14.1 | 18.6 | 122 | 7.45 |

**Table 5-5 Monte Carlo simulation results for coliform simulation with no light dependent die-off.**

| Site Name | No Light dependence Max | Avg | Min | Monte Carlo Most Prob. | Median | Max | Min |
|---|---|---|---|---|---|---|---|
| | Coliform/100 mL | | | Coliform/100 mL | | | |
| Park | 21.5 | 10.8 | 4.29 | 13.2 | 12.9 | 74.2 | 2.89 |
| Onyx | 25.6 | 15.5 | 8.95 | 17.8 | 18.6 | 98.0 | 5.64 |
| Ruby | 24.2 | 16.1 | 10.4 | 18.2 | 19.5 | 101 | 7.76 |
| Bayshore | 29.6 | 16.8 | 9.97 | 19.5 | 20.4 | 338 | 8.45 |
| ViaGenoa | 17.2 | 14.1 | 11.8 | 15.8 | 17.0 | 80.1 | 9.87 |
| 43rdStreet | 33.6 | 28.4 | 24.4 | 28.8 | 33.1 | 134 | 14.5 |
| 38th Street | 33.0 | 27.4 | 23.0 | 28.2 | 31.6 | 128 | 14.1 |
| 33rd Street | 27.9 | 24.7 | 21.4 | 25.7 | 28.8 | 99.1 | 13.3 |
| 19th St | 23.5 | 18.9 | 16.3 | 20.4 | 22.4 | 77.9 | 10.3 |
| 15th St | 18.7 | 15.1 | 12.7 | 17.0 | 17.8 | 68.6 | 10.2 |
| 10th St | 18.8 | 14.4 | 11.5 | 16.6 | 17.4 | 87.5 | 9.90 |
| Alvarado | 17.1 | 11.2 | 6.37 | 13.2 | 13.5 | 103 | 5.25 |
| Abalone | 17.9 | 8.53 | 3.14 | 11.5 | 10.2 | 65.2 | 2.17 |
| Sapphire | 15.7 | 8.23 | 3.45 | 11.2 | 10.0 | 79.9 | 2.75 |
| N Street Beach | 11.3 | 3.26 | 0.44 | 6.46 | 2.88 | 43.4 | 0.38 |
| Rocky Pt | 11.1 | 3.02 | 0.36 | 6.61 | 2.57 | 41.3 | 0.31 |
| Dunes | 30.4 | 23.4 | 18.7 | 23.4 | 26.9 | 457 | 16.6 |
| Dunes N | 30.1 | 21.8 | 16.9 | 21.9 | 25.7 | 512 | 14.9 |
| North Star Beach | 38.6 | 22.0 | 12.4 | 24.5 | 26.9 | 906 | 11.9 |
| Promontory Point | 15.9 | 14.5 | 13.0 | 14.8 | 16.6 | 49.0 | 9.62 |
| DeAnza Pier | 29.0 | 17.2 | 10.5 | 19.1 | 20.9 | 389 | 9.09 |
| Garnet Ave | 16.9 | 10.3 | 5.26 | 12.9 | 12.6 | 107 | 4.26 |
| Lids Island | 18.8 | 14.9 | 12.3 | 16.6 | 17.8 | 71.6 | 10.1 |
| Harbor Patrol Beach | 15.5 | 6.29 | 1.58 | 10.0 | 7.24 | 52.9 | 1.21 |
| Grand Canal | 34.1 | 16.9 | 8.29 | 18.6 | 20.0 | 154 | 4.79 |

**Table 5-6 Monte Carlo simulation results for conservative coliform simulation.**

| Site Name | Conservative Simulation Max | Avg | Min | Monte Carlo Most Prob. | Median | Max | Min |
|---|---|---|---|---|---|---|---|
| | Coliform/100 mL | | | Coliform/100 mL | | | |
| Park | 19.6 | 9.58 | 3.20 | 13.2 | 15.5 | 1682 | 0.368 |
| Onyx | 22.3 | 13.5 | 6.53 | 17.8 | 22.4 | 1984 | 0.715 |
| Ruby | 26.4 | 17.6 | 10.2 | 24.5 | 29.5 | 2382 | 1.10 |
| Bayshore | 48.7 | 28.9 | 13.6 | 37.2 | 46.8 | 4850 | 1.40 |
| ViaGenoa | 34.1 | 26.5 | 19.0 | 32.4 | 44.7 | 2990 | 2.01 |
| 43rdStreet | 34.9 | 29.0 | 22.9 | 43.7 | 49.0 | 2669 | 2.48 |
| 38th Street | 34.9 | 29.0 | 22.9 | 43.7 | 49.0 | 2670 | 2.47 |
| 33rd Street | 35.0 | 29.1 | 22.9 | 43.7 | 49.0 | 2669 | 2.47 |
| 19th St | 34.4 | 28.2 | 22.0 | 38.9 | 47.9 | 2656 | 2.64 |
| 15th St | 35.7 | 27.8 | 20.8 | 38.9 | 46.8 | 2821 | 2.40 |
| 10th St | 35.5 | 25.8 | 18.3 | 31.6 | 43.7 | 2972 | 1.97 |
| Alvarado | 30.0 | 18.9 | 8.27 | 26.9 | 30.9 | 2652 | 0.875 |
| Abalone | 20.5 | 8.42 | 2.34 | 12.3 | 13.2 | 1746 | 0.280 |
| Sapphire | 24.0 | 11.7 | 3.61 | 15.1 | 18.6 | 2127 | 0.405 |
| N Street Beach | 15.2 | 3.71 | 0.344 | 5.25 | 4.68 | 1285 | 0.040 |
| Rocky Pt | 14.6 | 3.32 | 0.268 | 4.79 | 3.98 | 1225 | 0.031 |
| Dunes | 76.6 | 55.6 | 34.2 | 56.2 | 91.2 | 8006 | 2.83 |
| Dunes N | 70.8 | 51.6 | 32.3 | 52.5 | 85.1 | 7373 | 2.55 |
| North Star Beach | 84.0 | 51.8 | 23.1 | 46.8 | 81.3 | 9040 | 2.03 |
| Promontory Point | 21.5 | 17.0 | 12.2 | 24.0 | 28.8 | 1889 | 1.31 |
| DeAnza Pier | 54.0 | 32.2 | 15.4 | 36.3 | 51.3 | 5500 | 1.55 |
| Garnet Ave | 28.6 | 16.6 | 6.31 | 20.4 | 26.9 | 2595 | 0.676 |
| Lids Island | 36.6 | 28.1 | 20.9 | 38.9 | 47.9 | 2846 | 2.42 |
| Harbor Patrol Beach | 17.9 | 6.35 | 1.21 | 10.5 | 9.55 | 1510 | 0.149 |
| Grand Canal | 21.4 | 11.1 | 4.51 | 13.5 | 18.2 | 1885 | 0.498 |

**Table 5-7 Monte Carlo simulation results for base coliphage stimulation.**

| Site Name | Base Simulation Max | Avg | Min | Monte Carlo Most Prob. | Median | Max | Min |
|---|---|---|---|---|---|---|---|
| | Coliphage/100 mL | | | Coliphage/100 mL | | | |
| Park | 0.317 | 0.163 | 0.060 | 0.004 | 0.047 | 188 | 0.0001 |
| Onyx | 0.345 | 0.221 | 0.115 | 0.007 | 0.063 | 185 | 0.0002 |
| Ruby | 0.394 | 0.270 | 0.169 | 0.009 | 0.076 | 197 | 0.0003 |
| Bayshore | 0.536 | 0.364 | 0.207 | 0.016 | 0.098 | 198 | 0.0005 |
| ViaGenoa | 0.483 | 0.368 | 0.283 | 0.011 | 0.102 | 225 | 0.0006 |
| 43rdStreet | 0.701 | 0.506 | 0.412 | 0.007 | 0.083 | 227 | 0.0004 |
| 38th Street | 0.692 | 0.497 | 0.403 | 0.007 | 0.085 | 228 | 0.0004 |
| 33rd Street | 0.669 | 0.483 | 0.396 | 0.007 | 0.085 | 227 | 0.0005 |
| 19th St | 0.581 | 0.441 | 0.354 | 0.008 | 0.123 | 262 | 0.0005 |
| 15th St | 0.560 | 0.419 | 0.327 | 0.009 | 0.126 | 295 | 0.0006 |
| 10th St | 0.522 | 0.377 | 0.277 | 0.010 | 0.115 | 273 | 0.0006 |
| Alvarado | 0.418 | 0.277 | 0.135 | 0.009 | 0.081 | 227 | 0.0003 |
| Abalone | 0.326 | 0.144 | 0.045 | 0.004 | 0.042 | 202 | 0.0001 |
| Sapphire | 0.350 | 0.184 | 0.064 | 0.006 | 0.052 | 192 | 0.0001 |
| N Street Beach | 0.237 | 0.064 | 0.007 | 0.003 | 0.013 | 139 | 0.0000 |
| Rocky Pt | 0.229 | 0.058 | 0.006 | 0.002 | 0.012 | 134 | 0.0000 |
| Dunes | 0.734 | 0.573 | 0.424 | 0.046 | 0.240 | 193 | 0.0018 |
| Dunes N | 0.694 | 0.535 | 0.399 | 0.037 | 0.158 | 203 | 0.0016 |
| North Star Beach | 0.763 | 0.516 | 0.302 | 0.041 | 0.117 | 185 | 0.0009 |
| Promontory Point | 0.378 | 0.292 | 0.235 | 0.006 | 0.065 | 148 | 0.0004 |
| DeAnza Pier | 0.570 | 0.388 | 0.228 | 0.017 | 0.100 | 195 | 0.0005 |
| Garnet Ave | 0.396 | 0.244 | 0.104 | 0.008 | 0.069 | 205 | 0.0002 |
| Lids Island | 0.572 | 0.423 | 0.327 | 0.010 | 0.135 | 320 | 0.0006 |
| Harbor Patrol Beach | 0.281 | 0.108 | 0.024 | 0.004 | 0.028 | 164 | 0.0000 |
| Grand Canal | 0.349 | 0.189 | 0.084 | 0.005 | 0.052 | 188 | 0.0002 |

**Table 5-8 Monte Carlo simulation results for coliphage simulation with no light dependent die-off.**

| | No Light dependence | | | Monte Carlo | | | |
|---|---|---|---|---|---|---|---|
| Site Name | Max | Avg | Min | Most Prob. | Median | Max | Min |
| | Coliphage/100 mL | | | Coliphage/100 mL | | | |
| Park | 0.384 | 0.200 | 0.070 | 0.008 | 0.052 | 204 | 0.0002 |
| Onyx | 0.418 | 0.275 | 0.135 | 0.011 | 0.071 | 207 | 0.0003 |
| Ruby | 0.474 | 0.340 | 0.203 | 0.015 | 0.087 | 224 | 0.0005 |
| Bayshore | 0.704 | 0.477 | 0.252 | 0.026 | 0.115 | 235 | 0.0006 |
| ViaGenoa | 0.576 | 0.471 | 0.347 | 0.019 | 0.120 | 259 | 0.0008 |
| 43rdStreet | 0.766 | 0.642 | 0.519 | 0.013 | 0.107 | 256 | 0.0008 |
| 38th Street | 0.759 | 0.632 | 0.507 | 0.013 | 0.110 | 257 | 0.0008 |
| 33rd Street | 0.730 | 0.617 | 0.501 | 0.013 | 0.110 | 257 | 0.0009 |
| 19th St | 0.646 | 0.561 | 0.446 | 0.015 | 0.145 | 310 | 0.0008 |
| 15th St | 0.638 | 0.532 | 0.409 | 0.017 | 0.148 | 337 | 0.0009 |
| 10th St | 0.606 | 0.480 | 0.342 | 0.017 | 0.132 | 314 | 0.0008 |
| Alvarado | 0.516 | 0.352 | 0.162 | 0.015 | 0.091 | 258 | 0.0004 |
| Abalone | 0.395 | 0.176 | 0.053 | 0.008 | 0.047 | 219 | 0.0001 |
| Sapphire | 0.430 | 0.230 | 0.075 | 0.010 | 0.060 | 213 | 0.0002 |
| N Street Beach | 0.292 | 0.078 | 0.008 | 0.005 | 0.015 | 152 | 0.0000 |
| Rocky Pt | 0.280 | 0.070 | 0.007 | 0.004 | 0.013 | 146 | 0.0000 |
| Dunes | 1.010 | 0.796 | 0.553 | 0.078 | 0.288 | 240 | 0.0025 |
| Dunes N | 0.918 | 0.736 | 0.508 | 0.066 | 0.195 | 258 | 0.0019 |
| North Star Beach | 1.015 | 0.705 | 0.381 | 0.056 | 0.148 | 228 | 0.0011 |
| Promontory Point | 0.440 | 0.368 | 0.281 | 0.011 | 0.078 | 167 | 0.0005 |
| DeAnza Pier | 0.754 | 0.513 | 0.279 | 0.030 | 0.120 | 234 | 0.0007 |
| Garnet Ave | 0.495 | 0.310 | 0.124 | 0.013 | 0.079 | 235 | 0.0003 |
| Lids Island | 0.657 | 0.535 | 0.407 | 0.017 | 0.151 | 369 | 0.0009 |
| Harbor Patrol Beach | 0.340 | 0.133 | 0.028 | 0.006 | 0.032 | 179 | 0.0001 |
| Grand Canal | 0.419 | 0.233 | 0.099 | 0.008 | 0.060 | 207 | 0.0002 |

**Table 5-9 Monte Carlo simulation results for conservative coliphage simulation.**

| | Conservative Simulation | | | Monte Carlo | | | |
|---|---|---|---|---|---|---|---|
| Site Name | Max | Avg | Min | Most Prob. | Median | Max | Min |
| | Coliphage/100 mL | | | Coliphage/100 mL | | | |
| Park | 0.562 | 0.298 | 0.098 | 0.015 | 0.093 | 249 | 0.0002 |
| Onyx | 0.640 | 0.415 | 0.194 | 0.022 | 0.132 | 266 | 0.0005 |
| Ruby | 0.725 | 0.519 | 0.294 | 0.027 | 0.162 | 292 | 0.0008 |
| Bayshore | 1.072 | 0.738 | 0.371 | 0.047 | 0.219 | 330 | 0.0010 |
| ViaGenoa | 0.956 | 0.769 | 0.534 | 0.035 | 0.240 | 363 | 0.0015 |
| 43rdStreet | 1.287 | 1.077 | 0.819 | 0.032 | 0.263 | 385 | 0.0017 |
| 38th Street | 1.277 | 1.064 | 0.802 | 0.032 | 0.263 | 386 | 0.0017 |
| 33rd Street | 1.246 | 1.048 | 0.798 | 0.032 | 0.263 | 386 | 0.0017 |
| 19th St | 1.148 | 0.965 | 0.722 | 0.035 | 0.309 | 473 | 0.0016 |
| 15th St | 1.088 | 0.883 | 0.637 | 0.035 | 0.295 | 484 | 0.0016 |
| 10th St | 0.994 | 0.773 | 0.520 | 0.033 | 0.257 | 440 | 0.0014 |
| Alvarado | 0.808 | 0.552 | 0.236 | 0.027 | 0.174 | 341 | 0.0006 |
| Abalone | 0.578 | 0.262 | 0.073 | 0.013 | 0.083 | 264 | 0.0002 |
| Sapphire | 0.651 | 0.352 | 0.108 | 0.021 | 0.110 | 275 | 0.0003 |
| N Street Beach | 0.433 | 0.116 | 0.011 | 0.009 | 0.028 | 187 | 0.0000 |
| Rocky Pt | 0.415 | 0.104 | 0.009 | 0.008 | 0.023 | 180 | 0.0000 |
| Dunes | 1.558 | 1.207 | 0.798 | 0.115 | 0.457 | 349 | 0.0030 |
| Dunes N | 1.426 | 1.123 | 0.736 | 0.098 | 0.331 | 369 | 0.0025 |
| North Star Beach | 1.541 | 1.073 | 0.559 | 0.091 | 0.275 | 325 | 0.0018 |
| Promontory Point | 0.702 | 0.577 | 0.418 | 0.026 | 0.158 | 238 | 0.0009 |
| DeAnza Pier | 1.143 | 0.793 | 0.412 | 0.054 | 0.229 | 330 | 0.0012 |
| Garnet Ave | 0.755 | 0.482 | 0.181 | 0.025 | 0.148 | 313 | 0.0005 |
| Lids Island | 1.104 | 0.883 | 0.636 | 0.035 | 0.302 | 518 | 0.0016 |
| Harbor Patrol Beach | 0.500 | 0.197 | 0.038 | 0.012 | 0.058 | 219 | 0.0001 |
| Grand Canal | 0.613 | 0.348 | 0.140 | 0.017 | 0.107 | 254 | 0.0004 |

## CHAPTER 6.0

# WATER QUALITY MODELING: ALTERNATIVES ANALYSIS

Coliphage simulations were performed for the following five alternatives (EOA Inc. et al. 2001):

1) Existing model with no input from San Diego Creek.
2) Existing model with no input from Santa Ana Delhi Channel.
3) Existing model with no input from Big Canyon Wash.
4) Existing model with decreased loading from vessels to 10% of its current level.
5) Existing model with decreased loading from bathers to 50% of its current level.

Each simulation was run for a 9-month period using a 13-day repeating tide and the same boundary conditions as the coliphage calibration, with the exception of the changed parameter for each alternative as noted above. The Back Bay Drain was not included in the alternative simulations, because dry weather flow was permanently diverted from the Bay in March 2001. Boundary conditions were held constant (or used the same weekly or hourly variation) throughout the 9-month period. Meteorological data from the Irvine CIMIS station were used from March through November 1998.

Alternative simulation results are plotted at the Dunes for each alternative scenario in Figures 6-1 through 6-5. Alternative 1 resulted in a decrease in concentrations at the Dunes of approximately 0.03 coliphage/100 mL. Alternative 2 concentrations decreased by about 0.01 coliphage/100 mL at the Dunes. Alternatives 3 and 5 produced no appreciable change from the base simulation concentrations. The largest concentration reductions resulted from Alternative 4 with reduced vessel loading. For this alternative, concentrations at the Dunes decreased by 0.05 to 0.1 coliphage/100 mL. This reduction is very sensitive to the large range in uncertainty in the vessel loading.

If the load reductions from the alternatives were combined, the concentration decreases would be cumulative. The water quality and health based implications for these alternative analyses are discussed in Section 8.4.

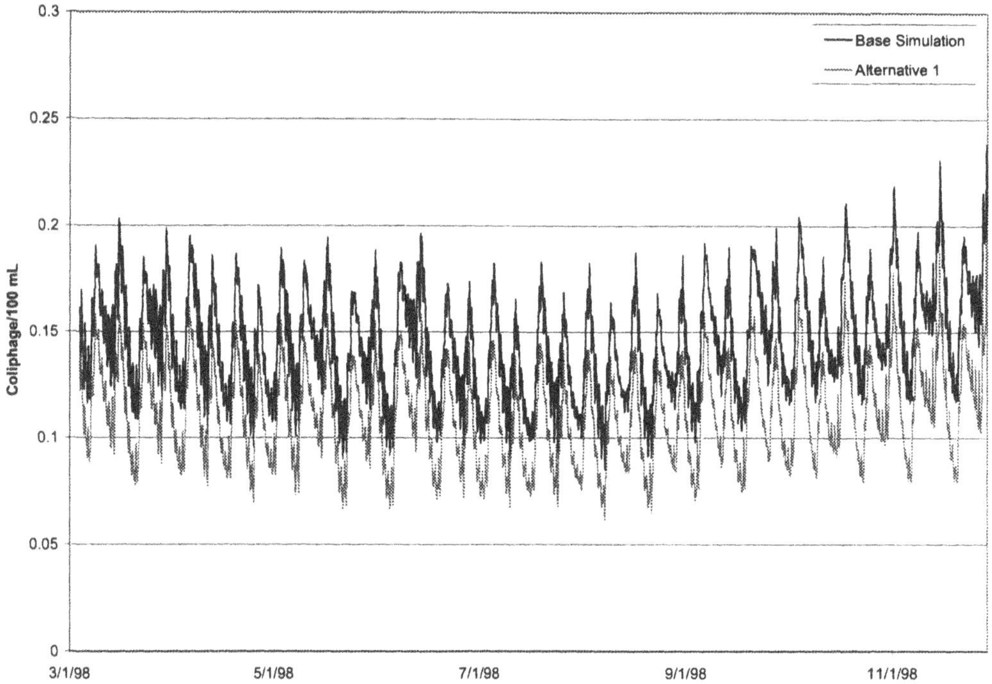


**Figure 6-1. Nine-month simulation results for Alternative 1 with base simulation results**

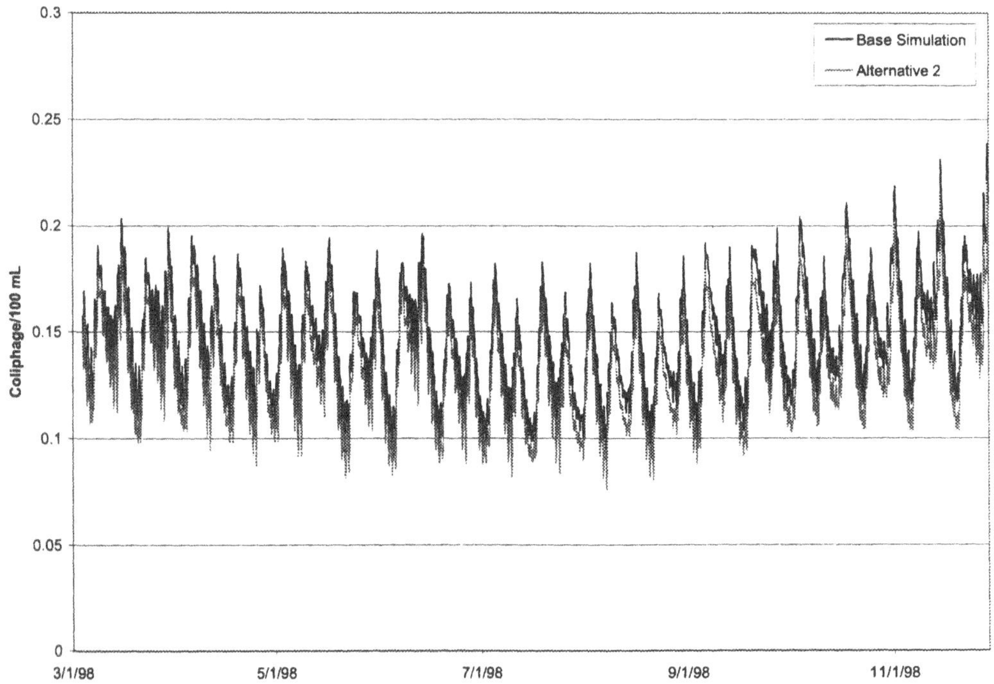


**Figure 6-2. Nine-month simulation results for Alternative 2 with base simulation results.**

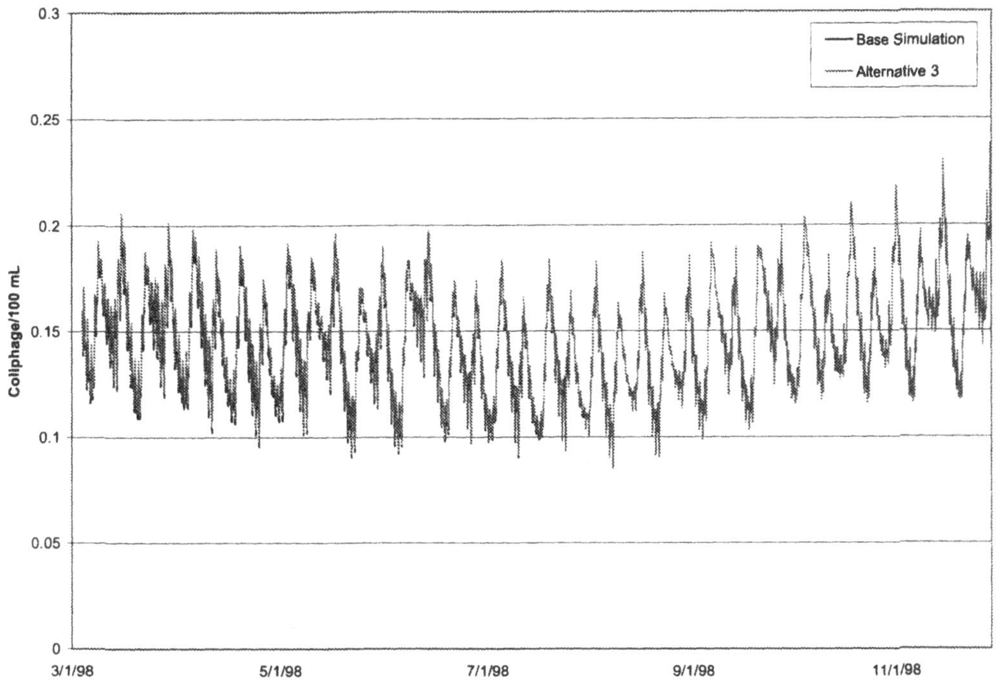


**Figure 6-3 . Nine-month simulation results for Alternative 3 with base simulation results.**

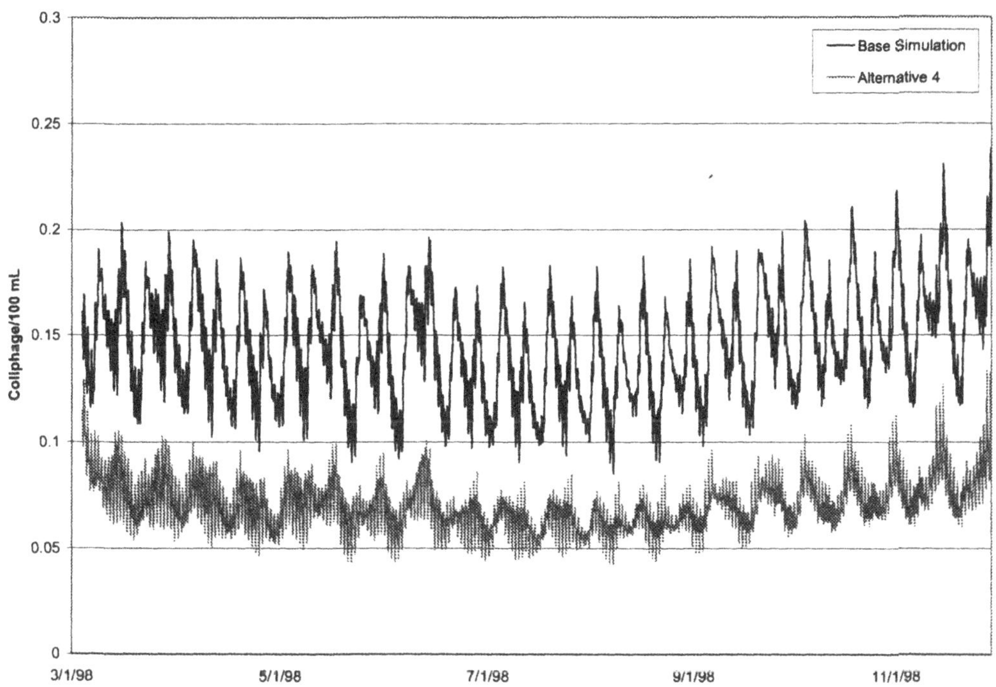


**Figure 6-4. Nine-month simulation results for Alternative 4 with base simulation results.**

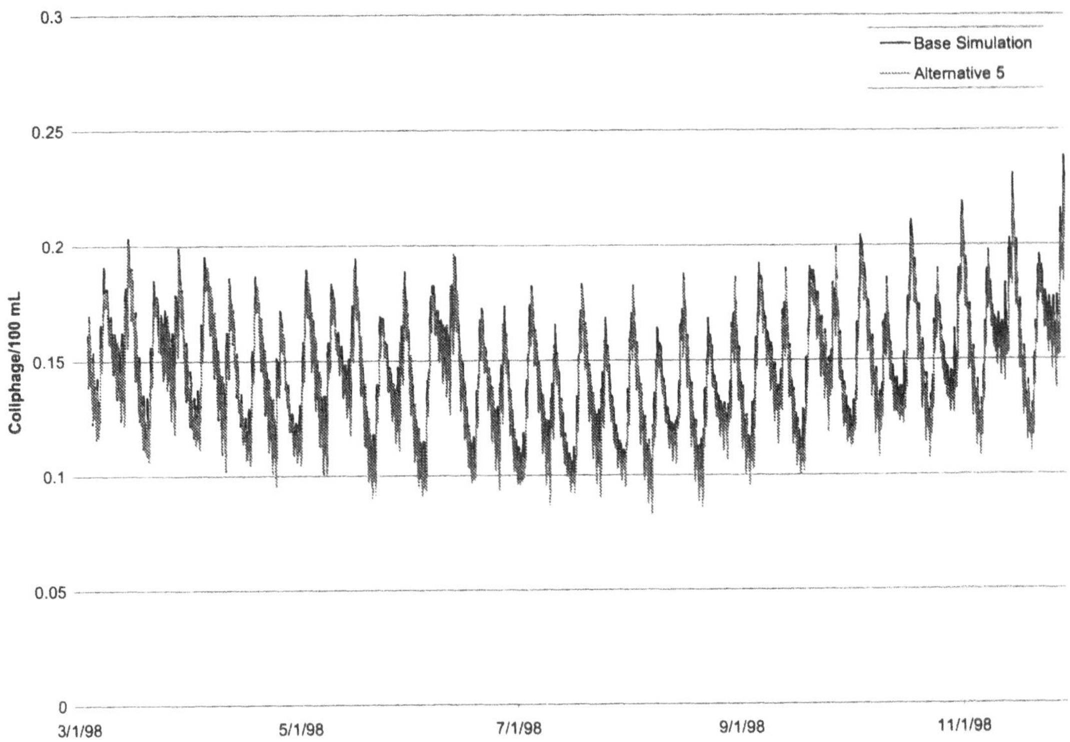


**Figure 6-5. Nine-month simulation results for Alternative 5 with base simulation results.**

# CHAPTER 7.0

# WATER QUALITY MODELING: SUMMARY

The RMA finite element model of the Newport Bay had been previously calibrated for simulation of hydrodynamics, sediment transport, and basic water quality parameters including salinity, temperature, dissolved oxygen, nutrients, phytoplankton, and macrophytes. For the current study, this proven model was applied to the transport of fecal coliform and coliphage.

The difficulty in fecal coliform and coliphage modeling is the uncertainty in loading rates. For the fecal coliform calibration, best estimates were made based on available data for loads from the creeks, monitored storm drains, and survey values for vessel loads and bather loads. Literature values were used to estimate die-off rates in darkness and sunlight. To bring computed concentrations into agreement with observed data, loading estimates were made for the small, unmonitored storm drains in the lower Bay, and a distributed load was applied to represent undocumented non-point sources. This calibration method produced computed coliform concentrations that are in reasonable agreement with observed data throughout most of the Bay. At the Dunes East site, the known loads were not sufficient to produce fecal coliform concentrations as high as those observed. It is possible that die-off rates from the literature are not applicable to Newport Bay or that the intermittent flow to Back Bay Drain is responsible for the discrepancy. If lower die-off rates could be justified for use in the model, concentrations in the Dunes would increase while concentrations throughout the rest of the Bay could be kept near currently calibrated levels by reducing the distributed load.

Nearly all of the coliphage data collected for this investigation were reported to be below detectable limits (1 pfu/100 mL). Results from model simulations based on best estimates of loading, and estimating die-off rates from literature values show concentrations throughout the Bay less than 1 coliphage/100 mL. Given the limited amount of quantifiable coliphage data, a rigorous calibration was not possible. However, if more quantifiable data should become available in the future, a more rigorous calibration could be carried out.

To investigate the importance of individual loads, a post-processing routine was developed to determine the contribution from individual loads to total coliform and coliphage concentrations at 25 locations throughout the Bay. Individual load simulations were performed for each constituent using: (1) die-off rates from the respective calibration/configuration simulations, (2) no light dependent die-off, and (3) no die-off. The distributed load was found to be the largest contributor to total coliform concentrations using the calibrated die-off rates. With no die-off and the distributed load eliminated, San Diego Creek becomes the largest contributor to total coliform concentrations. For the coliphage simulations, the vessel loading was the most important load, regardless of the die-off rate because it was the largest load in the vicinity of the 25 sites that were examined.

To examine the importance of the uncertainty in the input data, Monte Carlo uncertainty analyses were performed using the simulations described above. These analyses resulted in concentration probability distributions at 25 locations throughout the Bay based on the probability distributions of data and uncertainty in die-off rates. The Monte Carlo results indicated that for coliform simulations, the maximum expected concentration ranges at each site were up to 4 times the maximum computed values for the base simulation, up to 24 times the maximum computed values for the no light dependent die-off simulation, and up to 108 times the computed values for the conservative simulation. The large uncertainty in the vessel coliphage load dominates the maximum expected concentration range calculated in the Monte Carlo uncertainty analysis. The maximum coliphage concentration ranges were up to 600 times the maximum computed value.

With no light dependent die-off, the most probable coliform values were increased by up to 3.6 times the base simulation and most probable coliphage concentrations were increased by up to 2 times the base simulation. With no die-off, most probable coliform concentrations for the conservative simulation were increased by up to 8.5 times the base simulation, and the most probable coliphage concentrations were increased by up to 4.7 times the base simulation.

Five alternative scenarios were simulated. Results indicated that elimination of San Diego Creek coliphage loads would reduce concentrations at the 25 sites from 6% to 30% on average, and elimination of Santa Ana Delhi channel coliphage loads would reduce concentrations at the 25 sites by 2% to 11% on average. Reduction of bather loading by 50% and elimination of the Big Canyon Wash coliphage load would have no significant impact on concentrations. The largest impact was from reduction of the vessel loads to 10% of the current level. This alternative resulted in average reductions ranging from 50% to 80% at the 25 sites. If the load reductions from the alternatives were combined, the concentration decreases would be cumulative.

## CHAPTER 8.0

# HEALTH RISK ASSESSMENT: RESULTS

### 8.1 Characterization of Existing Risk: Disease Transmission Modeling

The number of individuals in each of the epidemiological states (refer to Figure 1-2) at the end of 1000 disease transmission model simulations (each simulation represents 270 days) was found to be very similar for background and background plus body contact recreation exposure conditions (Figure 8-1). It is also noteworthy that (1) the number of individuals in the Post-Infection (P) state was similar to the number in the Susceptible (S) state, and (2) the numbers of individuals in the Carrier (C) and Diseased (D) states were substantially lower than those in states S and P.

To estimate the attributable risk posed by body contact recreation in Newport Bay above background exposure, the number of individuals in the diseased state due to background exposure at the end of each simulation was subtracted from those in the diseased state from background plus REC-1 exposure. This difference accounts for the number of individuals in the diseased state that is attributable to REC-1 exposure including subsequent secondary transmission of disease. The results of this analysis indicate that ~99.9% of the population in the diseased state at any time is due to background exposure compared to ~0.1% due to recreational contact (Figure 8-2). It should be noted, however, that all members of the population are subject to background exposure, whereas only those who choose to recreate in the Bay are subject to the incremental REC-1 exposure.

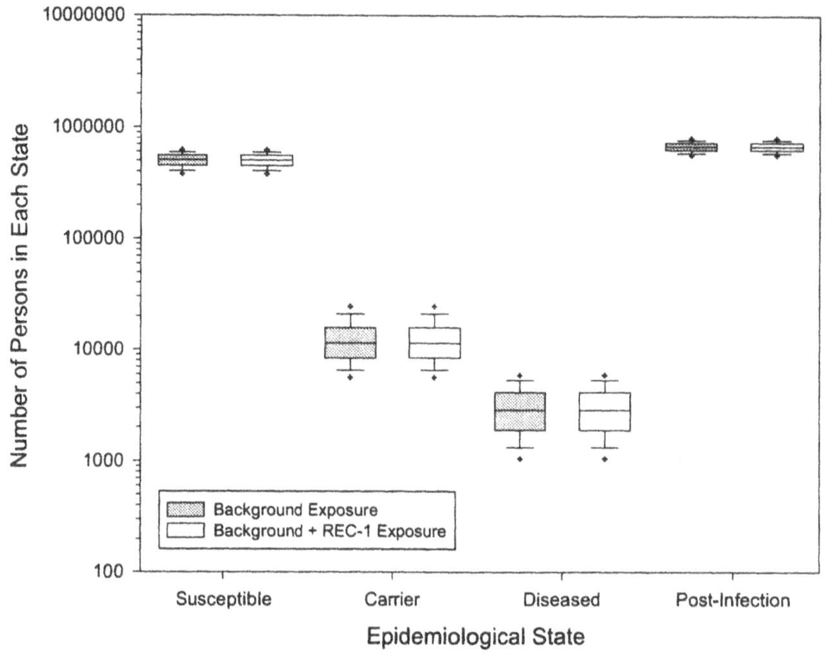


**Figure 8-1. Disease Transmission Model Output for Background and Background Plus REC-1 Exposure**

The applicable water quality criteria for REC-1 exposure are all derived directly or indirectly from EPA's best estimate of the accepted illness rates for the fecal coliform criteria (U.S. EPA 1986), which are 8 illnesses per 1,000 swimmers for fresh water beaches and 19 illnesses per 1,000 swimmers at marine beaches. To compare the output from the disease transmission model to those levels, the following approximation was employed:

$$\text{Risk of disease per swim event} = \Delta P * \text{sigma}_{dp} / N_s \qquad (8.1\text{-}1)$$

where
$\Delta P$ = Disease Prevalence attributable to REC-1 exposure
$\text{sigma}_{dp}$ = Rate that Diseased individuals move to Post infection state
$N_s$ = Average number of swimmers per day.

Converting the population-based attributable risk value to individual-based equivalents using the above formula results in a measure of risk that is comparable to appropriate benchmark risk values. With this conversion,the entire distribution of predicted diseased cases per swimming event fell below EPA's accepted median illness per swim event ratio for both marine and fresh waters (U.S. EPA 1986; U.S. EPA 2000) (Figure 8-3).

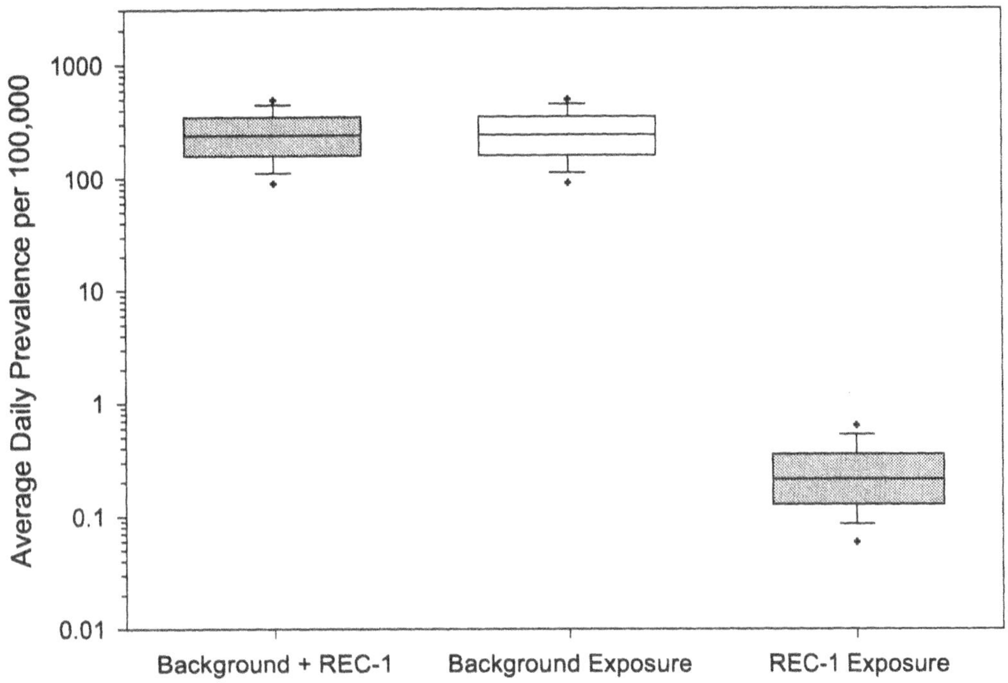


**Figure 8-2. Average Daily Prevalence in the Newport Bay Watershed Attributable to Background and REC-1 Exposures**

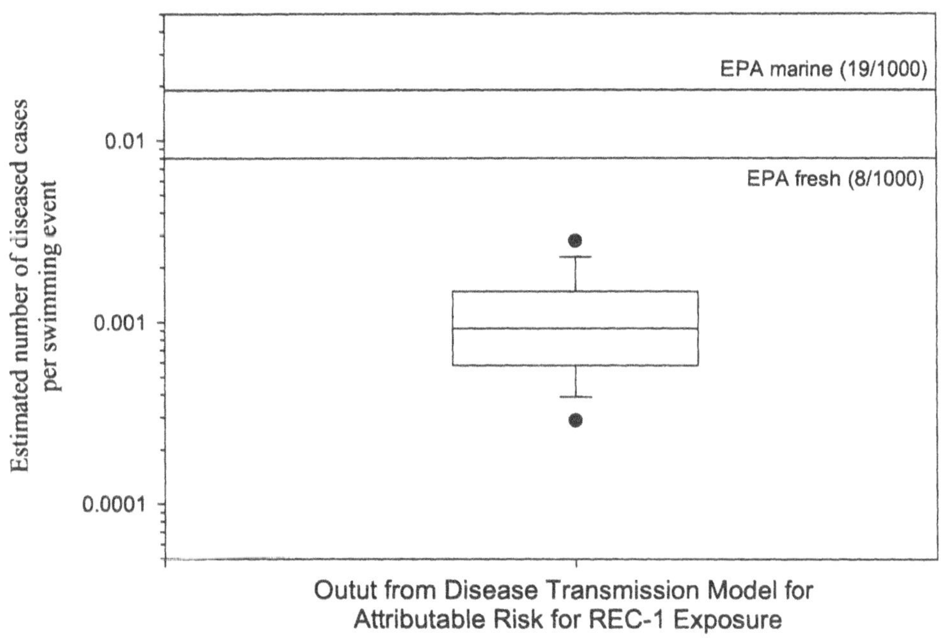


**Figure 8-3. Comparison of Risk of Disease per Swimming Event with EPA Benchmark Levels**

## 8.2 Characterization of Existing Risk: Enterococcus Data

To corroborate the disease transmission modeling results, a separate analytical approach was employed to evaluate enterococcus data collected in the most heavily used recreation site in Newport Bay. This approach involved applying a traditional risk framework (National Research Council 1983) along with EPA's recommended dose-response function for enterococcus (U.S. EPA 1986). A total of 246 enterococcus observations were collected and analyzed at the specified site during the recreation seasons of 1999 and 2000. These enterococcus data were assumed to be representative of the bacterial water quality at this recreational site during that time period; 48% of those data were reported to be below the detection limit of 10 MPN/100 mL.

The enterococcus data were fit to a log-normal distribution via the method of maximum likelihood. Based on the resultant distribution and EPA's equation relating enterococci density in marine water to illness rate (U.S. EPA 1986), Monte Carlo simulations were used to estimate the expected distribution of disease attributable to REC-1 contact at this recreational site. Results from these simulations were compared with the output from the disease transmission modeling efforts (Figure 8-4) and indicate that (1) the levels of disease predicted by both the enterococcus data and the disease transmission model are below the EPA's accepted levels for marine water, and (2) the levels of disease attributable to body contact recreation estimated by the disease transmission model are somewhat lower than those estimated by using enterococcus data. Several possible explanations for this second observation follow:

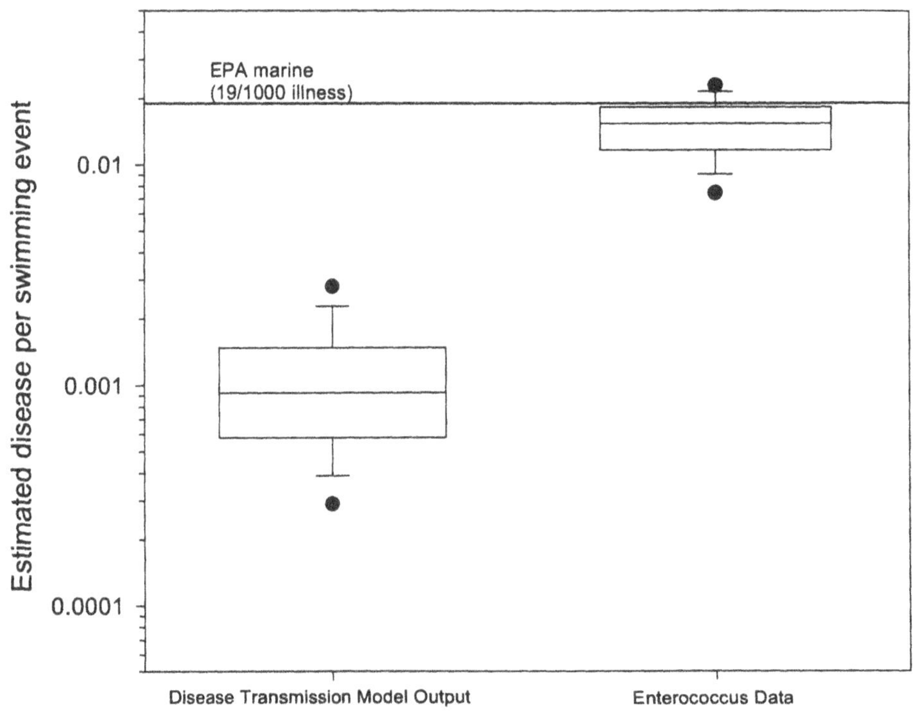


**Figure 8-4. Comparison of Disease Transmission Model Output with Estimated Risk Based on Enterococcus Data**

- The importance of a post-infective state (state P) for enteric viruses may not be sufficiently captured in EPA's concentration-response function for enterococcus;
- ~48% of the enterococcus observations used in the analysis were reported to be below the detection limit of 10 colonies per 100mL. Those observations were assumed to be present at a concentration equal to the detection limit. If a more rigorous approach, such as probit analysis were used to estimate the distribution of enterococcus concentrations, the resultant estimate of risk would be lower than that presented;
- The uncertainty associated with the enterococcus analysis is not captured in a manner commensurate with the uncertainty embedded in the disease transmission model output. For example, the enterococcus data employed in the analysis only represent the reported concentrations, not the 95% confidence intervals about the reported values. For the membrane filtration method, a reported concentration of 10 colonies per 100mL has upper and lower 95% confidence limits of 4.7 and 18.4 colonies per 100mL, respectively (APHA et al. 1995). Incorporating this uncertainty into the analysis would have the effect of increasing the variability about the reported median; and
- Use of enterococcus as an indicator organism for human health risk from body contact recreation may not be appropriate in this particular watershed for assessing risk from recreational contact: (1) It is possible that the enterococcus sources to Newport Bay are more abundant than sources contributing viruses relative to those in the studies used to generate the enterococcus concentration-response relation; and (2) A reanalysis of the data supporting the U.S. federal bacteriological water quality criteria for enterococcus in marine waters with similar salinity to Newport Bay, found significant differences in the mathematical relationship between swimming-associated gastroenteritis and enterococcus density at the two locations used in the study (Fleisher, 1991).

## 8.3 Sensitivity and Uncertainty Analysis

Sensitivity and uncertainty analyses were carried out on the disease transmission modeling results. The purpose of these analyses was to determine if output from the disease transmission modeling was highly sensitive to any of the input parameters that were known to be highly uncertain and/or variable, and to characterize the sensitivity of the results in terms of implications to risk management. The components of these studies included investigating: (1) The impact of increased concentrations of enteric viruses in the Bay; (2) The impact of increased viral loading from vessels; and (3) The impact of increased viral loading from bathers.

Concentration levels of viruses in Newport Bay were increased above those used previously by factors of 10, 100, and 10,000. These levels were selected to bracket conservatively the unlikely but possible underestimation of enteric viruses in Newport Bay due to analytical inefficiencies, variability in the ratio of male specific coliphage to viruses, and other factors that may not be accounted for. Sets of simulations were run using those new concentration levels along with the same parameter sets as those described previously. The results of this analysis indicate that even if the estimated concentrations of enteric viruses in Newport Bay were several orders of magnitude higher than they were estimated to be in this investigation, the predicted number of diseased cases per swimming event would be within the level of risk considered acceptable by EPA for recreational activities (Figure 8-5). A likely explanation for the results of this analysis derives directly from the shape of the dose-response function for rotavirus, in which the probability of infection increases with the log of concentration and is linear at low concentrations (Regli et al., 1991). At concentrations similar to those in this investigation, the probability of infection increases linearly with log increases in dose. Figure 8-5 indicates that the increase in the number of disease cases per swimming event with increased concentration of viruses in the Bay is consistent with an increase in probability of infection from the dose-response function.

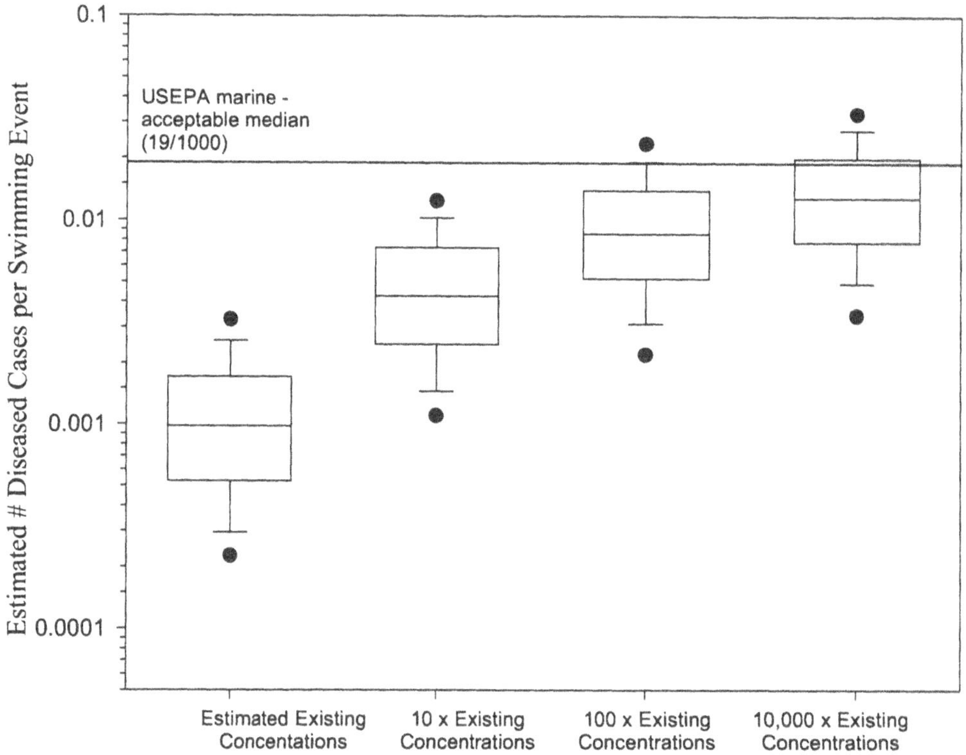


**Figure 8-5. Estimated Number of Diseased Cases per Swimming Event for Increased Levels of Enteric Viruses in Newport Bay**

The loading contribution from vessels was increased by a factor of 100 over that specified previously, and that from recreators by a factor of 200. The vessel waste loading estimate is based on a number of factors, including an estimate of $10^8$ infective virus particles per gram of feces and an estimated 10% of the vessels discharging waste to the Bay (EOA Inc. et al. 2001). The increase in loading from vessels investigated in this analysis would be consistent with increasing the estimate of infective virus particles per gram of feces to $10^{10}$ infective virus particles per gram (Flewett and Woode 1978) or assuming that 100% of the boats in the Bay are discharging waste. The increase in loading from bathers investigated in this analysis corresponds to the $99^{th}$ percentile of the distribution describing the estimated loading from bathers (EOA Inc. et al., 2001). The results of this analysis indicate that even if the levels of viruses contributed to Newport Bay via vessel waste loading were 100 times higher than estimated, or if bather loading were 200 times higher than estimated, the estimated risk to an individual per swimming event would still be within the levels considered acceptable by EPA (Figure 8-6).

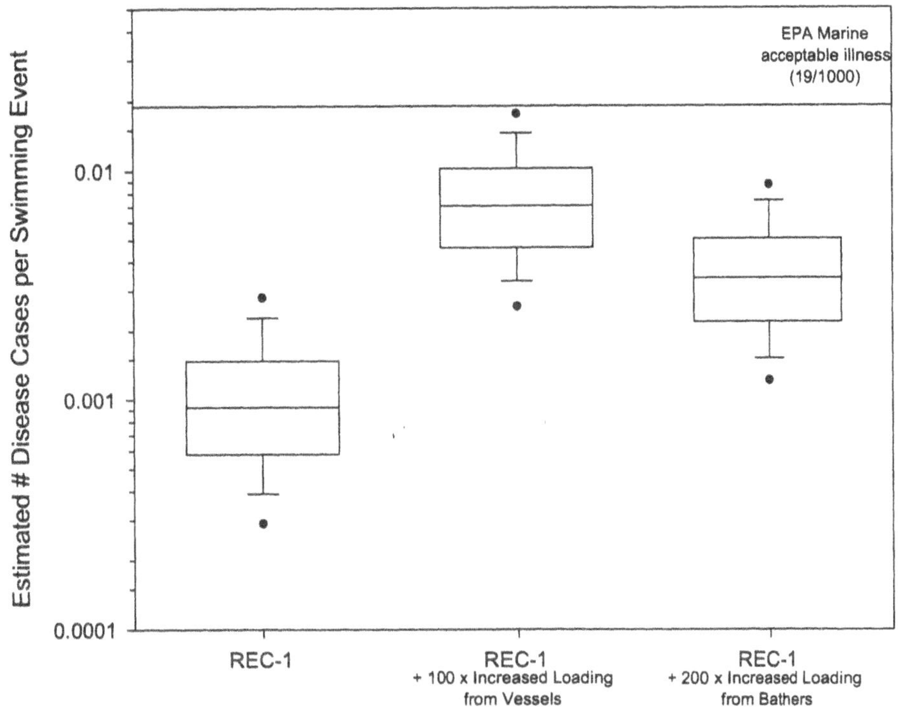


**Figure 8-6. Estimated Risk per Swim Event for REC-1 Exposure With and Without Increased Loadings**

## 8.4 Comparative Assessment and Planning Level Costs

Alternative management scenarios are theoretical or potential options for reducing the viral loading to Newport Bay. Six alternative management scenarios were developed and simulated employing the water quality and disease transmission models. Each alternative scenario involves reduction in virus loading to the Bay through the reduction or elimination of a specific source of pathogens. The first four alternatives focus on point source loadings to the Upper Bay such as storm drains and creeks. The last two alternatives focus on bathers and boaters as sources of pathogen loading to the Bay.

The scenarios considered represented existing conditions (Alternative 0), existing conditions with dry weather flow from one storm drain diverted (Alternative 1), Alternative 1 plus diverting dry weather viral loading from San Diego Creek (Alternative 2), Santa Ana Delhi (Alternative 3), and the Upper Bay storm drains (Alternative 4), and reducing viral loading from vessels by 90% (Alternative 5) and recreators by 50% (Alternative 6).

The results of the alternative analyses indicate the median estimated number of diseased cases per swim event is reduced from ~0.9 to 0.6 per 1000 swimmers under alternative 1. Implementation of Alternatives 2, 3, 4, or 6 would result in an incremental reduction of 1 case of disease per 10,000 swim events or less. Under Alternative 5, the median estimated number of diseased cases per swim event decreases from approximately 6 in 10,000 to approximately 3 in 10,000 (Figure 8-7).

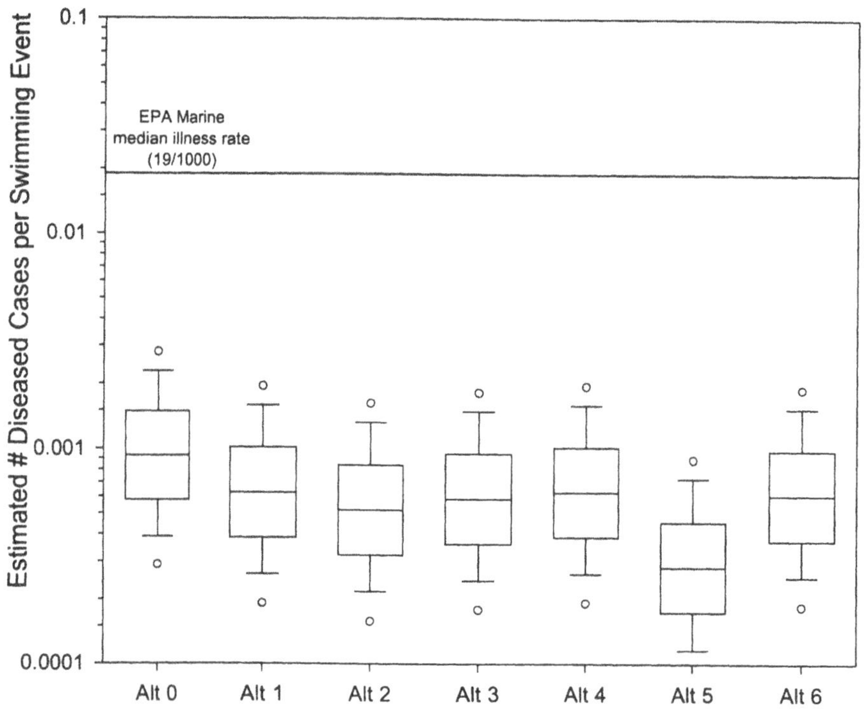


**Figure 8-7. Estimated Number of Diseased Cases per Swim Event for Existing Conditions and Alternative Scenarios**

Planning level cost estimates were developed for the alterative scenarios considered (Table 8-1). Planning level costs associated with Alternatives 1, 2, 3, and 4 involved estimated costs to treat or divert to the sanitary sewer Bay inflows. Planning level costs associated with Alternatives 5 and 6 involved estimating costs for implementing new public outreach programs (EOA Inc. et al. 2001). When the results of the alternative analyses are considered along with planning level costs, the following observations become apparent: Eliminating theoretical dry weather viral loading from San Diego Creek, Santa Ana Delhi, or Big Canyon Wash to Upper Newport Bay would be more expensive than other options and would not substantially reduce the existing risk to swimmers; and reducing pathogen loading from vessels may be more effective in terms of avoided disease cases per swim event than the other alternatives considered.

**Table 8-1. Planning Level Costs for Comparative Assessment Alternatives**

| Alternative | Summary | Estimated Annual Cost ($) | Probability of Achieving Goal |
|---|---|---|---|
| 0 | Existing conditions | - | |
| 1 | Back Bay storm drain diverted | $15,000[1,3] | |
| 2 | Alt.1 and water treatment for San Diego Creek | $2,359,000[1] | High[2] |
| 3 | Alt 1 and water treatment for Santa Ana Delhi | $1,405,000[1] | High[2] |
| 4 | Alt 1 and Upper Bay storm drains diverted to sanitary sewer | $219,000[1,4] | High[2] |
| 5 | Alt 1 and loading from vessels decreased to 10% estimated level | $25,000 - $70,000[5] | Unknown |
| 6 | Alt 1 and loading from bathers decreased to 50% estimated level | $25,000 - $50,000[5] | Unknown |

[1] Annualized capital cost is based on a design life of 20 years and a return rate of 10%
[2] Equivalent to disinfected tertiary effluent
[3] Currently in place
[4] Assumes drain diversion to sanitary sewer rather than treatment option; Assumes Community Outreach Campaign

# CHAPTER 9.0

# Health Risk Assessment: Discussion and Conclusions

## 9.1 Discussion

The California Regional Water Quality Control Board, Santa Ana Region identified and listed Newport Bay as a water quality limited receiving water body in accordance with Section 303(d) of the Clean Water Act. The 303(d) List identifies pathogens as a stressor of water quality impairment and indicates that urban runoff and/or storm sewers are primary sources ofpathogens. A TMDL for fecal coliform was adopted to ensure the reasonable protection of the Bay's beneficial uses (CA RWQCB 1999).

Based on the reported exceedances of the fecal coliform water quality criteria for body contact recreation and the results of the health risk assessment investigation, it is clear that interpreting the impairment of the REC-1 beneficial use in Newport Bay requires a more rigorous and comprehensive health risk evaluation than thc pass/fail tcst that has cvolved to regulate the quality of recreational waters. Such an approach would be consistent with (1) the basic principles of public health engineering regarding the use of sanitary surveys to identify and control potential sources to the maximum extent practicable, and (2) a health-based monitoring approach for recreational waters recently outlined by the World Health Organization (WHO) (WHO 1999). The WHO approach is to classify health risk as a function of both degree of overall fecal contamination and susceptibility to human contamination. In a similar manner, EPA's strategy for setting water quality criteria (U.S. EPA 1986; U.S. EPA 2000) was to vary the maximum allowable density based on the level of recreational use that a particular site receives (i.e., designated beaches have lower allowable maximum values than moderately, lightly, or infrequently used areas).

In this investigation, a risk assessment approach was taken to address the question of beneficial use impairment. If sufficient site-specific data or knowledge are available however, a less comprehensive approach may be suitable to address the issue of use impairment. For example, data from a sanitary survey, relevant water quality data, or specific knowledge about microbiological sources may allow a more qualitative approach to be employed. A site-by-site summary is provided in Table 9-1 demonstrating how the information presented herein may be used to facilitate a more qualitative interpretation of the Basin Plan Objective for fecal coliform with respect to risk to public health, and thus the impact on beneficial uses. The summary provided in Table 9-1 uses observed fecal coliform concentrations, levels of exposure, and seasonality to derive a relative level of public health concern associated with each of the monitoring sites. Risk managers in other watersheds may use a similar approach to that shown in Table 9-1 to consider the potential impairment of the REC-1 beneficial use within their watersheds if sufficient data or knowledge are available to support the use of a more qualitative approach.

**Table 9-1. Summary of Relative Public Health Concern Related to Fecal Coliform Exceedance in Newport**

| Site Name | Recreational Use Level [1] | Exceedances of FC WQC During Dry Season [2] | | Level of Public Health Concern Dry Season [7] | Exceedances of FC WQC During Wet Season [3] | | Level of Public Health Concern Wet Season [7] | Potential Future Controls [4] | | | |
|---|---|---|---|---|---|---|---|---|---|---|---|
| | | Log Mean | 90th %ile | | Log Mean | 90th %ile | | Vessel Waste | Swimmer Loading | Tributary Inflow | Others [9] |
| Vaughn's Launch | Use Prohibited | | √ | Very Low | √ | √ | Low | - | - | - | - |
| Ski Zone | | | | Very Low | | √ | Very Low | - | - | - | - |
| Big Canyon Creek | | | √ | Very Low | √ | √ | Low | - | - | - | - |
| San Diego Creek - Campus Dr. | | | √ | Very Low | √ | √ | Low | - | - | - | - |
| Backbay Drive Pipe [5] | | | √ | Very Low | | √ | Very Low | - | - | - | - |
| Santa Ana Delhi Channel | | √ | √ | Low | √ | √ | Low | - | - | - | - |
| 43rd Street Beach | Low Use | √ | √ | Medium | | √ | Low | √ | - | - | Flushing |
| 38th Street Beach | | | √ | Low | | √ | Low | - | - | - | Flushing |
| 33rd Street Channel | | | √ | Low | | √ | Low | - | - | - | Flushing |
| Rhine Channel | | | | Very Low | | √ | Low | - | - | - | Flushing |
| Sapphire Avenue Beach | | | | Very Low | | | Very Low | - | - | - | - |
| N Street Beach | | | | Very Low | | √ | Low | - | - | - | - |
| Promontory Point Channel | | | | Very Low | | | Very Low | - | - | - | - |
| De Anza | | | | Very Low | | √ | Low | - | - | - | - |
| Grand Canal | | | | Very Low | | | Very Low | - | - | - | - |
| Newport Blvd. Bridge | | | √ | Low | | √ | Low | - | - | - | Flushing |
| Park Avenue Beach | Medium Use | | | Very Low | | | Very Low | - | - | - | - |
| Onyx Avenue Beach | | | | Very Low | | √ | Low | - | - | - | - |
| Ruby Avenue Beach | | | | Very Low | | √ | Low | - | - | - | - |
| Bayshore Beach [6] | | | √ | Medium | | √ | Low | √ | - | - | - |
| Via Genoa Beach | | | | Very Low | | √ | Low | - | - | - | - |
| 15th Street Beach | | | √ | Low | | √ | Low | - | - | - | - |
| 10th Street Beach | | | | Very Low | | √ | Low | - | - | - | - |
| Alvarado/ Bay Isle Beach | | | | Very Low | | √ | Low | - | - | - | - |
| Abalone Avenue Beach | | | | Very Low | | √ | Low | - | - | - | - |
| Garnet Avenue Beach | | | | Very Low | | √ | Low | - | - | - | - |
| Harbor Patrol Beach | | | | Very Low | | √ | Low | - | - | - | - |
| 19th Street Beach | High Use | | | Very Low | | √ | Low | - | - | - | - |
| Rocky Point Beach | | | | Very Low | | | Very Low | - | - | - | - |
| North Star Beach | | | | Very Low | | √ | Low | - | - | - | - |
| Lido Yacht Club Beach | | | | Very Low | | √ | Low | - | - | - | - |
| Newport Dunes - Middle | Extra High Use | | | Low | | √ | Medium | √ | √ | - | Flushing |
| Newport Dunes - West | | | | Low | | √ | Medium | √ | √ | - | Flushing |
| Newport Dunes - North [8] | | | √ | High | | √ | Medium | √ | √ | - | Flushing |
| Newport Dunes - East | | | | Low | | √ | Medium | √ | √ | - | Flushing |

1. Use Levels are based on data collected as part of this investigatic
2. Dry Season defined as May through September
3. Wet Season defined as October through April
4. Potential structural changes or management options
5. Dry weather flow is now diverted to sanitary sewer
6. High Use on holidays
7. Level of Public Health Concern is derived as shown to the right
8. Exceedances may have been linked to flow from the Backbay Drive Pipe, see note 5.
9. Other potential control measures not investigated as part of the HRA

Level of Public Health Concern
Dry Season

| Indicator Concentration | Exposure: None | Low | Medium | High | Extra high |
|---|---|---|---|---|---|
| Low | Very Low | Very Low | Very Low | Very Low | Low |
| Medium | Very Low | Low | Low | Medium | High |
| High | Low | Medium | Medium | High | High |

Level of Public Health Concern
Wet Season

| Indicator Concentration | Exposure: None | Low | Medium | High | Extra high |
|---|---|---|---|---|---|
| Low | Very Low | Very Low | Very Low | Very Low | Low |
| Medium | Very Low | Low | Low | Low | Medium |
| High | Low | Low | Low | Medium | High |

Note: The relative level of Public Health Concern shown in these tables is based on the combination of exposure level, indicator concentration, basic principles of sanitary engineering, and professional judgement. Concern at a given site changes seasonally, with the relative use that the site receives.

## 9.2 Conclusions

### 9.2.1 303(d) Listing of Newport Bay for Fecal Coliform

The public health risk assessment investigation indicates that the exceedances of the Basin Plan fecal coliform objectives for the REC-1 beneficial use are temporally sporadic, geographically limited, and generally occur during the time of the year when REC-1 use is low and/or in areas of the Bay where the level of body contact recreation is low or is prohibited.

### 9.2.2 Relative Risk to Public Health from REC-1 Contact in Newport Bay

Given that the health risk assessment included a number of health protective assumptions, it is extremely likely that the estimated risk levels presented herein are higher than the "actual" levels of risk encountered by the population. Nevertheless, the risk of illness from REC-1 use in Newport Bay was estimated using two separate methods, both of which found that the estimated risk from REC-1 use was generally below levels considered tolerable by U.S. EPA for recreation in marine waters (U.S. EPA 1986; U.S. EPA 2000).

### 9.2.3 Controllable Sources

Urban runoff and/or storm sewers are identified in the 303(d) listing as likely sources of pathogens in Newport Bay. However, this investigation indicates that neither of those sources substantially impact the risk to public health from body contact recreation in Newport Bay during the time of the year that the vast majority of the recreation occurs.

Viral loadings from vessel sanitary waste and from swimmers were also investigated as controllable sources of pathogens in Newport Bay. Based on the estimated costs to implement a public outreach program and the estimated benefit in terms of reducing enteric virus disease, the reduction of vessel sanitary waste may be a control measure worth further consideration if risk managers feel that reduction of public health risk below the current levels is warranted.

### 9.2.4 Beneficial Use Impairment Evaluation

The fundamental question motivating investigations during the first phase of the Newport Bay fecal coliform TMDL was: "Do occasional exceedances of the fecal coliform objectives for body contact recreation necessarily imply an impairment to the beneficial use?". The results of the health risk assessment investigation indicate that the answer to this question is a resounding "no". The health risk assessment results clearly demonstrate that exposure is inextricably linked with the potential impairment of beneficial uses and must be considered in conjunction with water quality data to assess impairment of the beneficial use.

In summary, interpreting the impairment of the REC-1 beneficial use in Newport Bay requires a more rigorous and comprehensive health based evaluation approach than the pass/fail test that has evolved to regulate the quality of recreational waters. Risk managers in other watersheds may use a similar approach to consider the potential impairment of the REC-1 beneficial use within their watersheds. By considering potential levels of exposure and seasonality in conjunction with water quality data, the reasonable protection of the REC-1 beneficial use may be addressed in a streamlined manner emphasizing public health protection while concurrently balancing the costs of additional control measures.

# APPENDIX A

# COMPUTED AND OBSERVED FECAL COLIFORM CONCENTRATIONS AT NEWPORT BAY MONITORING SITES

Computed and observed fecal coliform concentrations at Ski Zone.

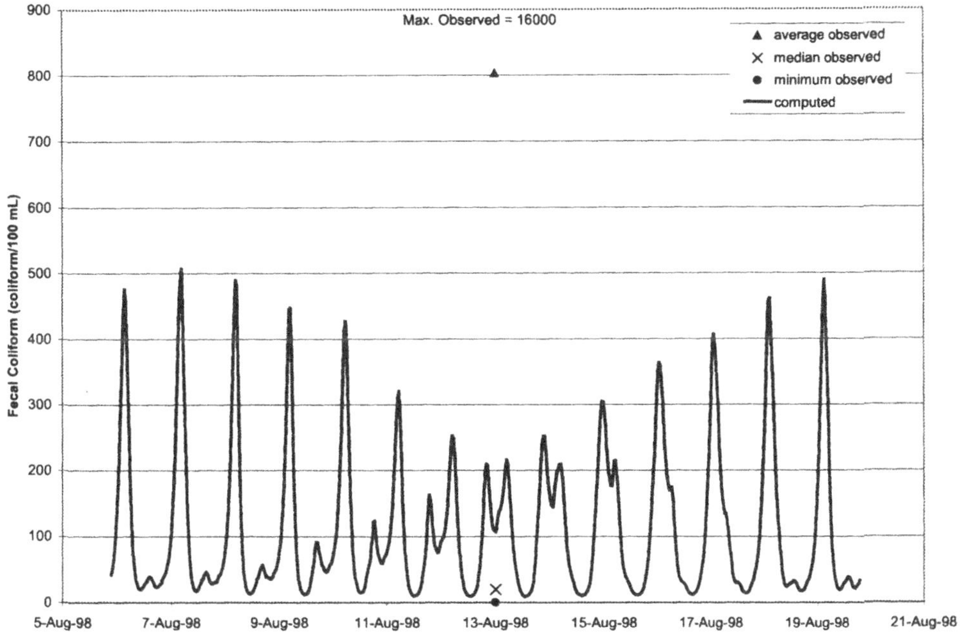


Computed and observed fecal coliform concentrations at Vaughn's Launch.

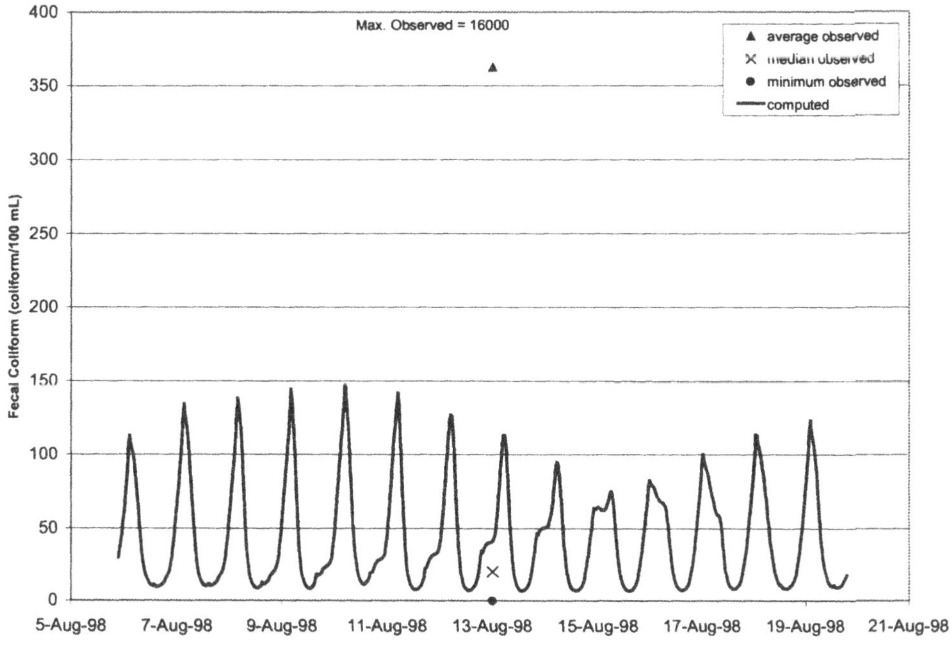

Computed and observed fecal coliform concentrations at North Star Beach.

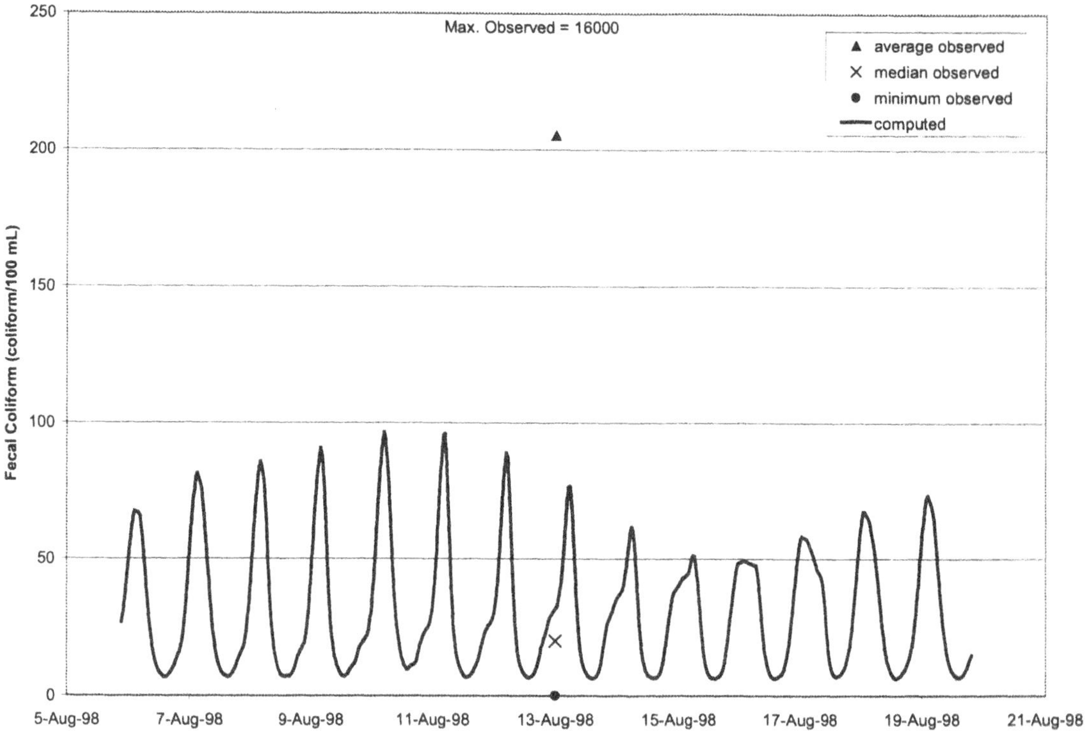


Computed and observed fecal coliform concentrations at DeAnza Pier.

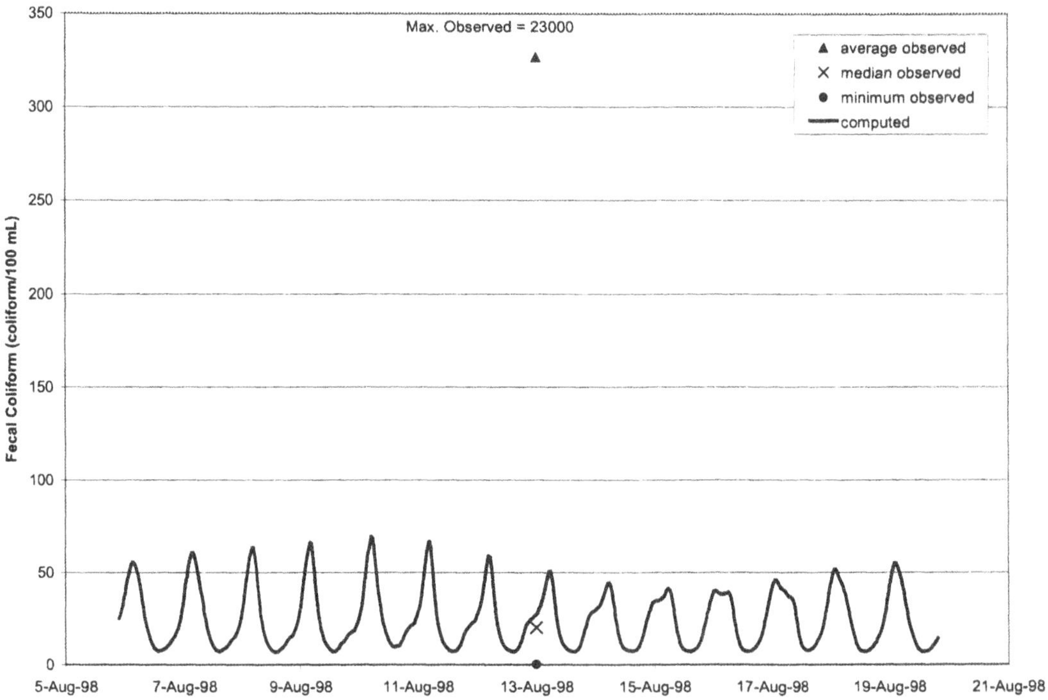

Computed and observed fecal coliform concentrations at Bayshore Beach.

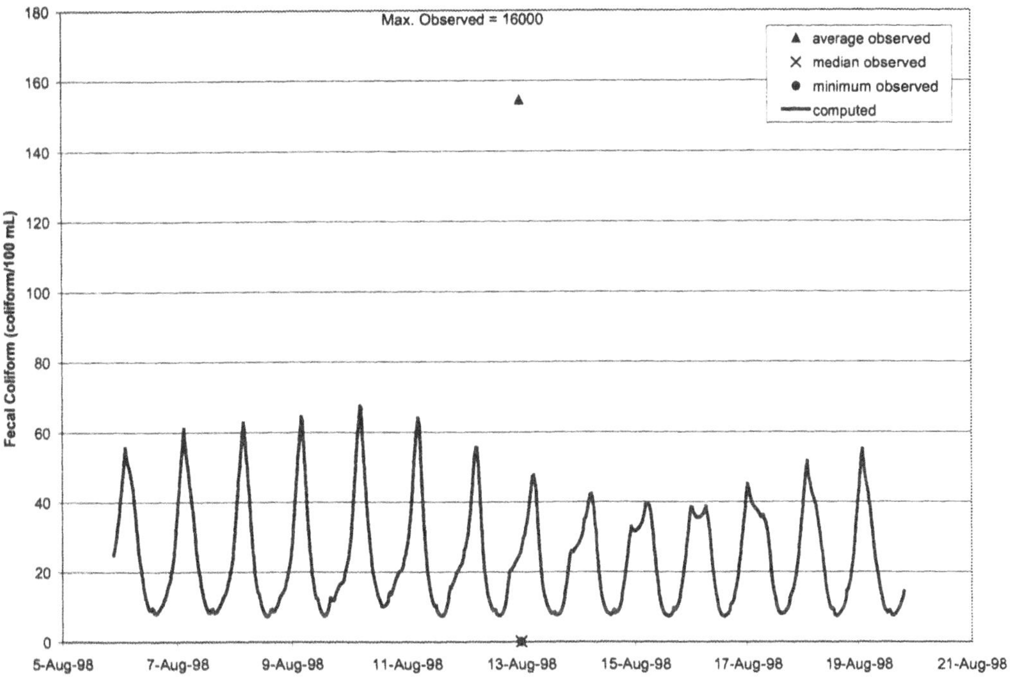


Computed and observed fecal coliform concentrations at Dunes North.

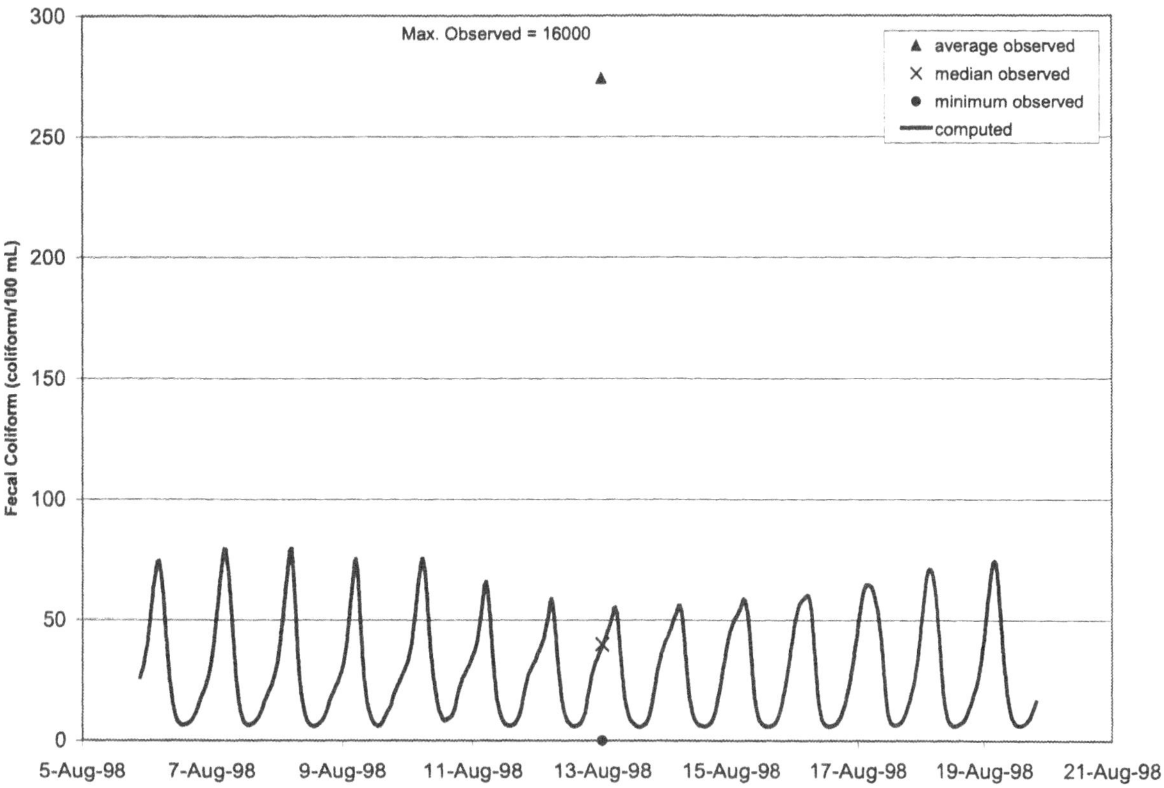

Computed and observed fecal coliform concentrations at Dunes East.

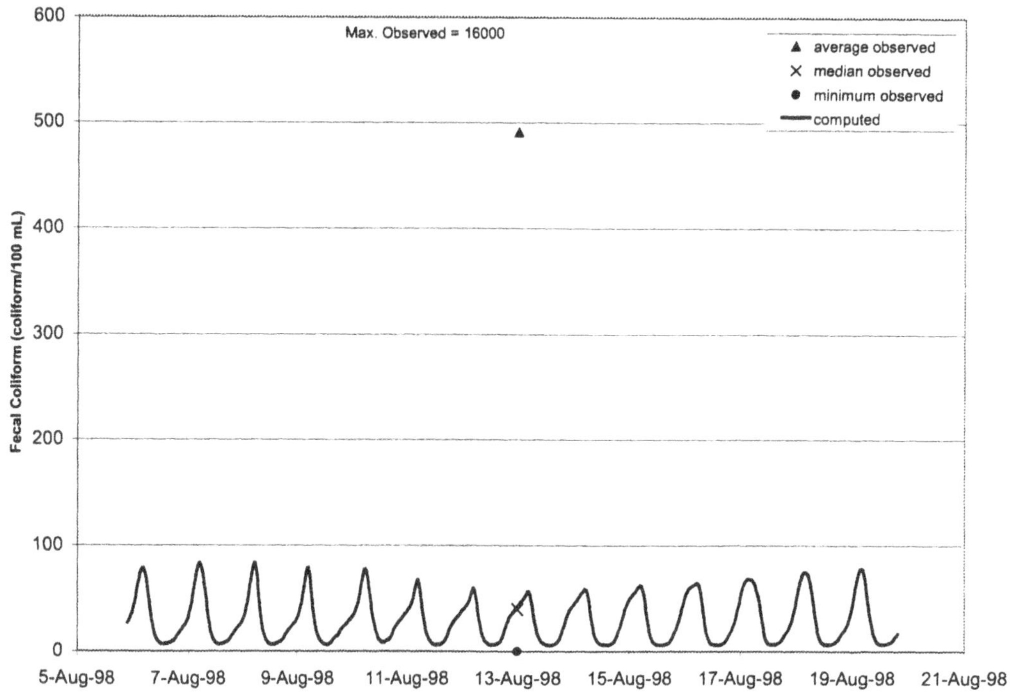


Computed and observed fecal coliform concentrations at Dunes West.

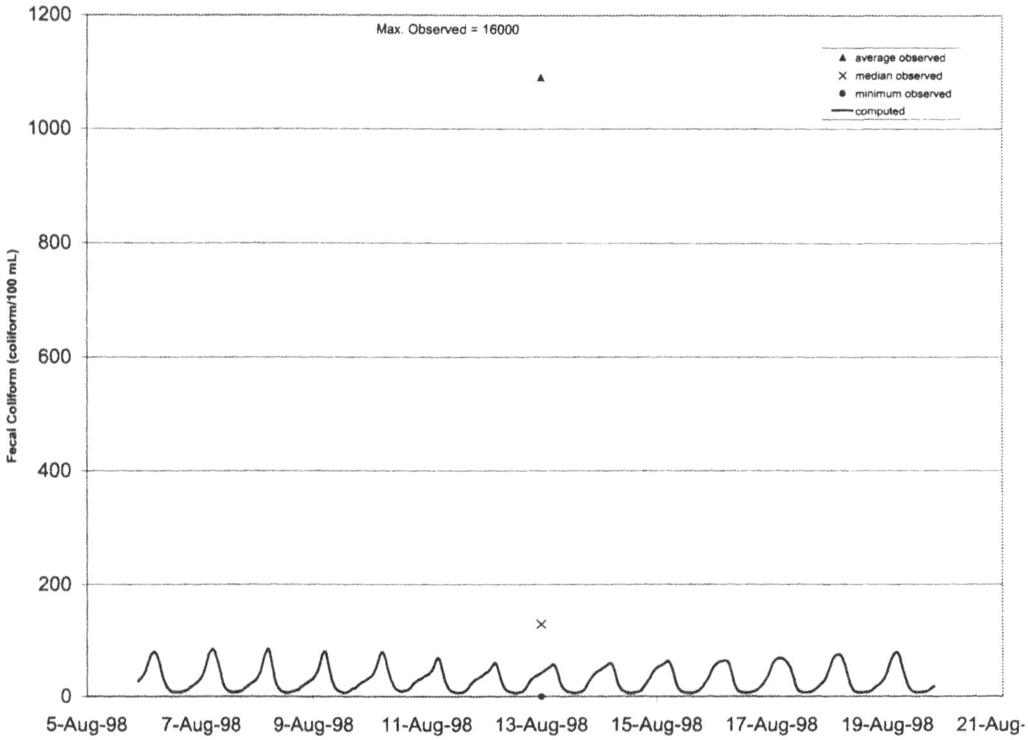

Computed and observed fecal coliform concentrations at Dunes Middle.

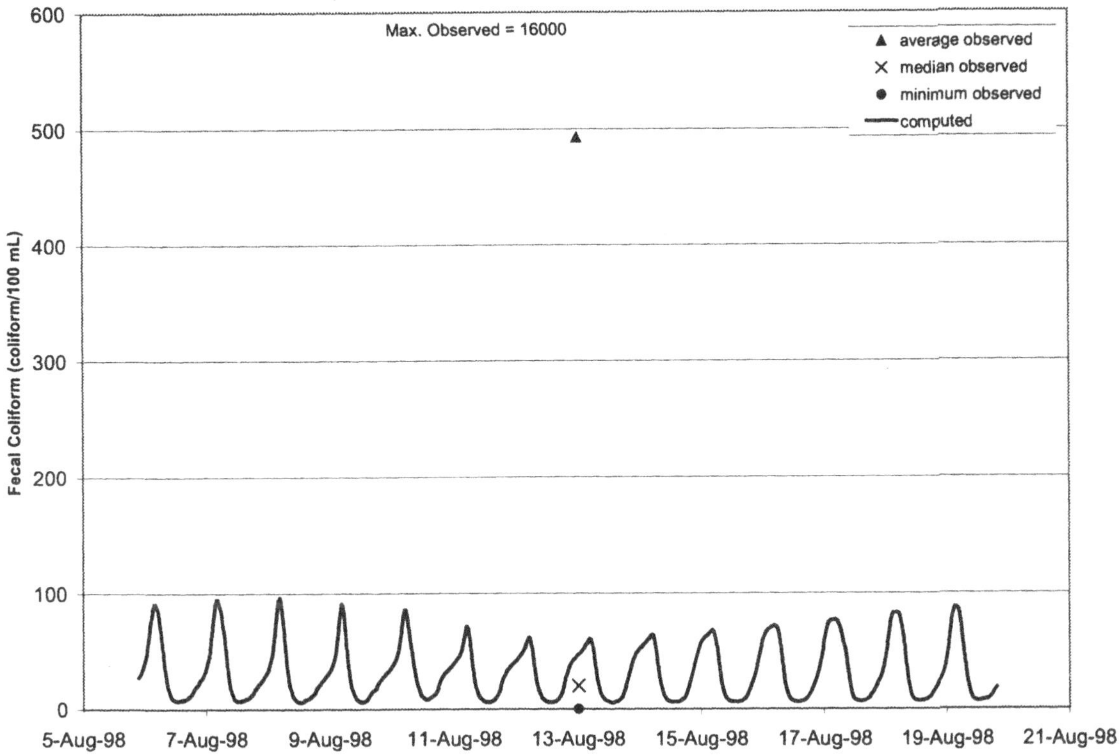


Computed and observed fecal coliform concentrations at Promontory Point.

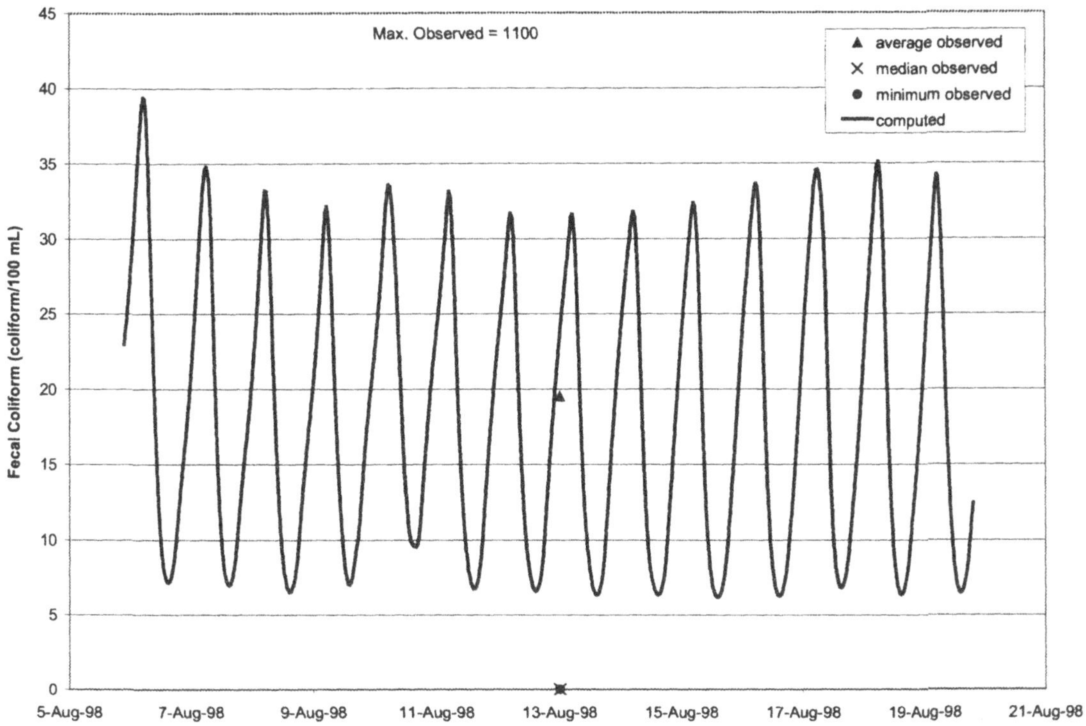

Computed and observed fecal coliform concentrations at Ruby Avenue.

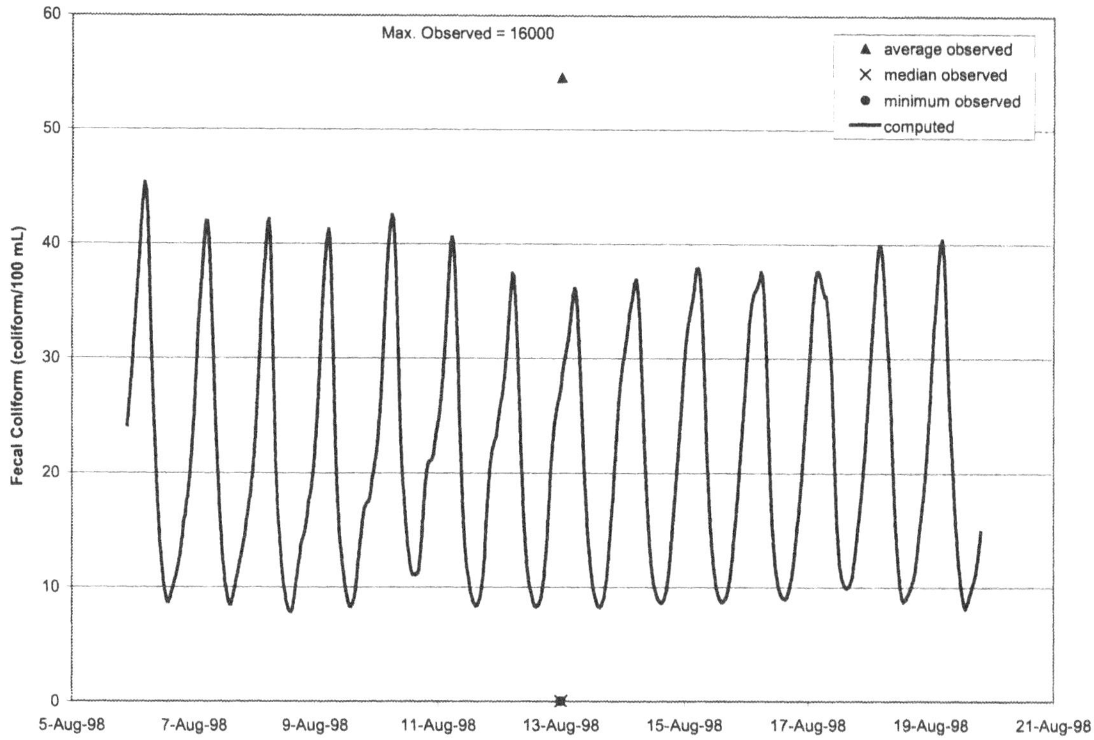


Computed and observed fecal coliform concentrations at Onyx Avenue.

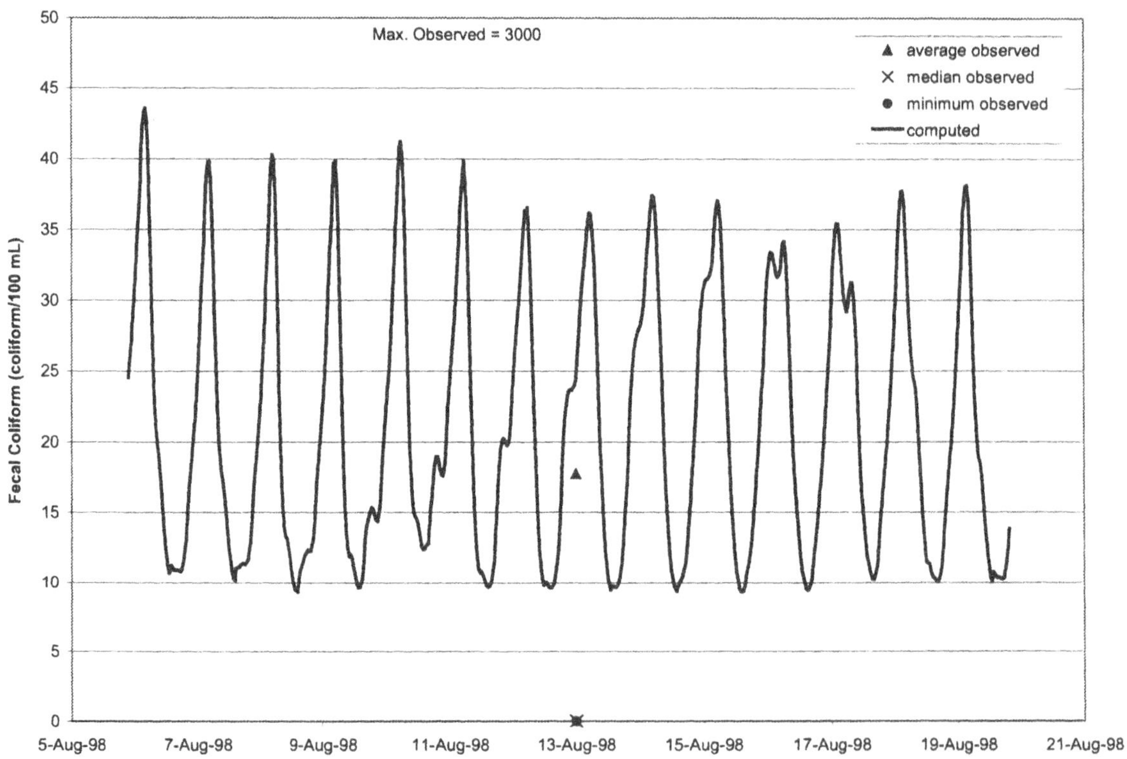

Computed and observed fecal coliform concentrations at Grand Canal

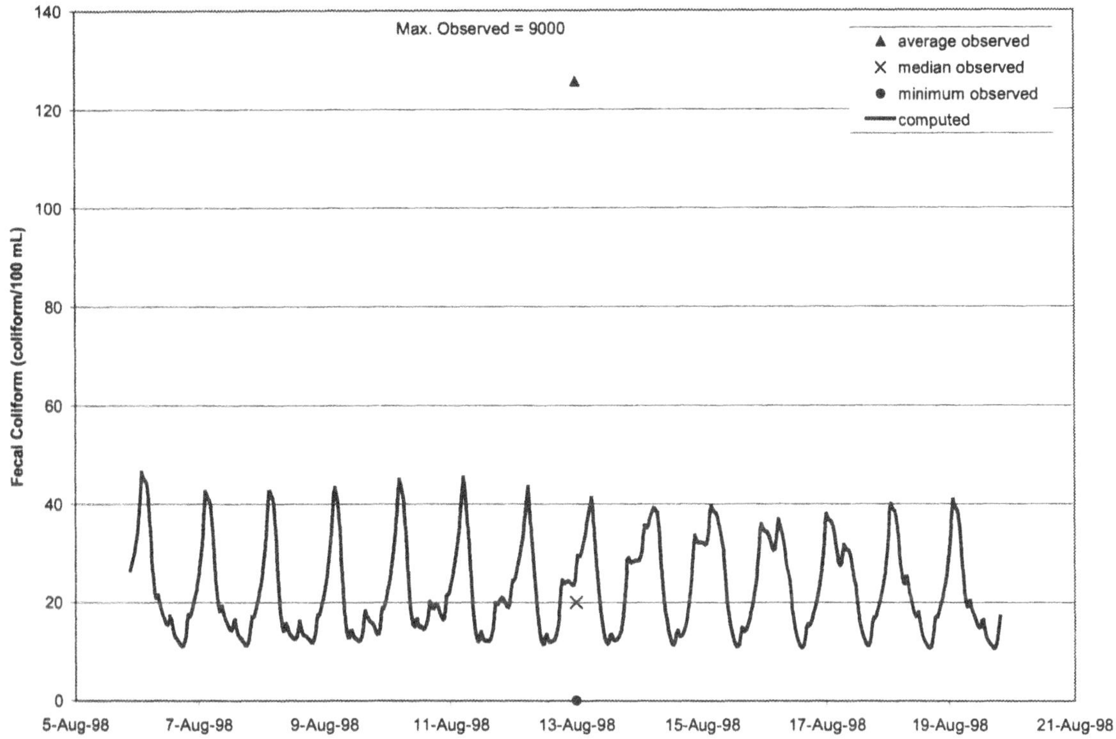


Computed and observed fecal coliform concentrations at Park Avenue

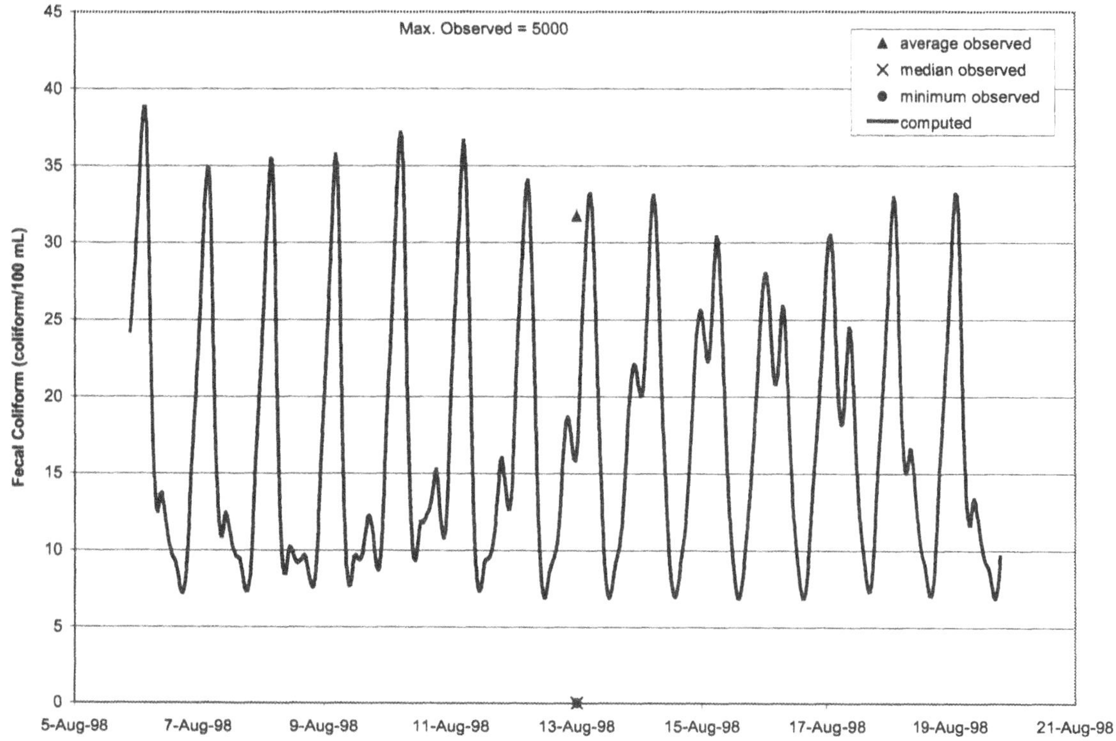

Computed and observed fecal coliform concentrations at Abalone Avenue

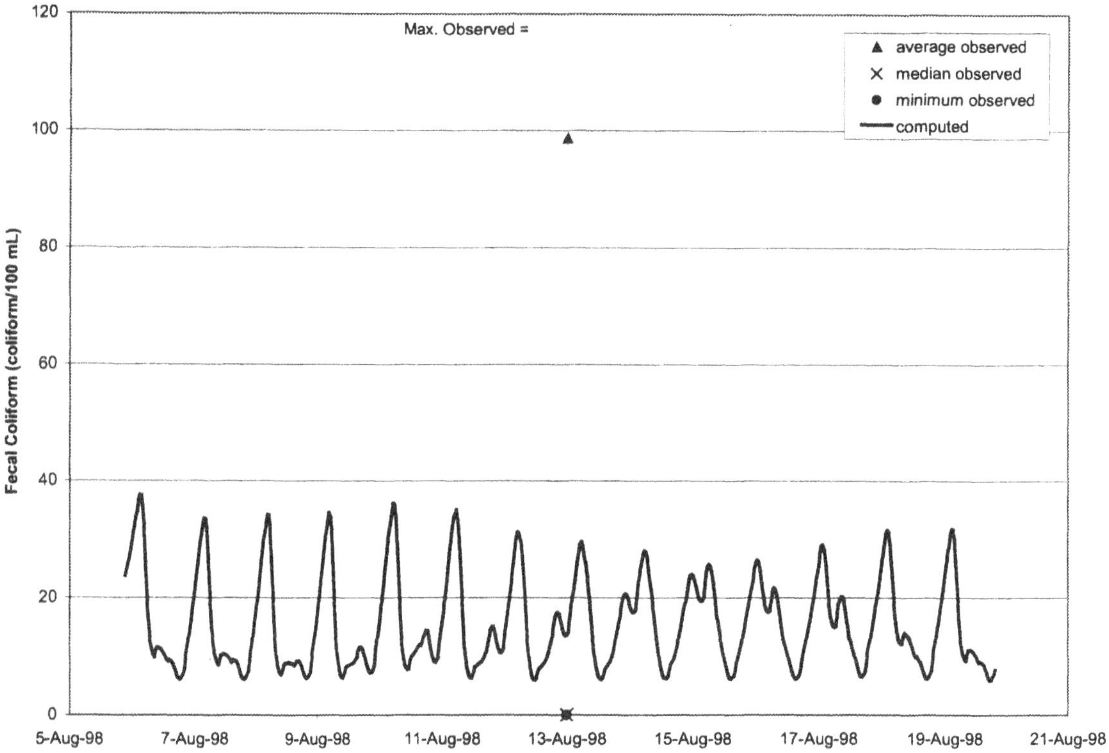


Computed and observed fecal coliform concentrations at Sapphire Avenue

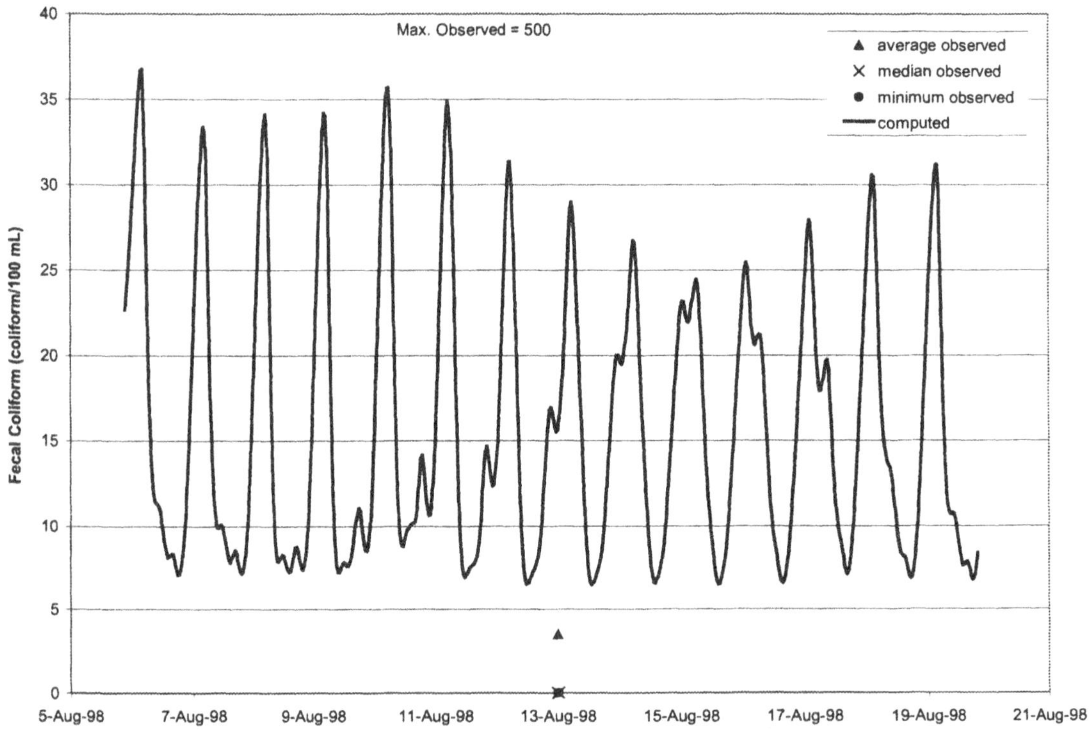

Computed and observed fecal coliform concentrations at Garnet Avenue.

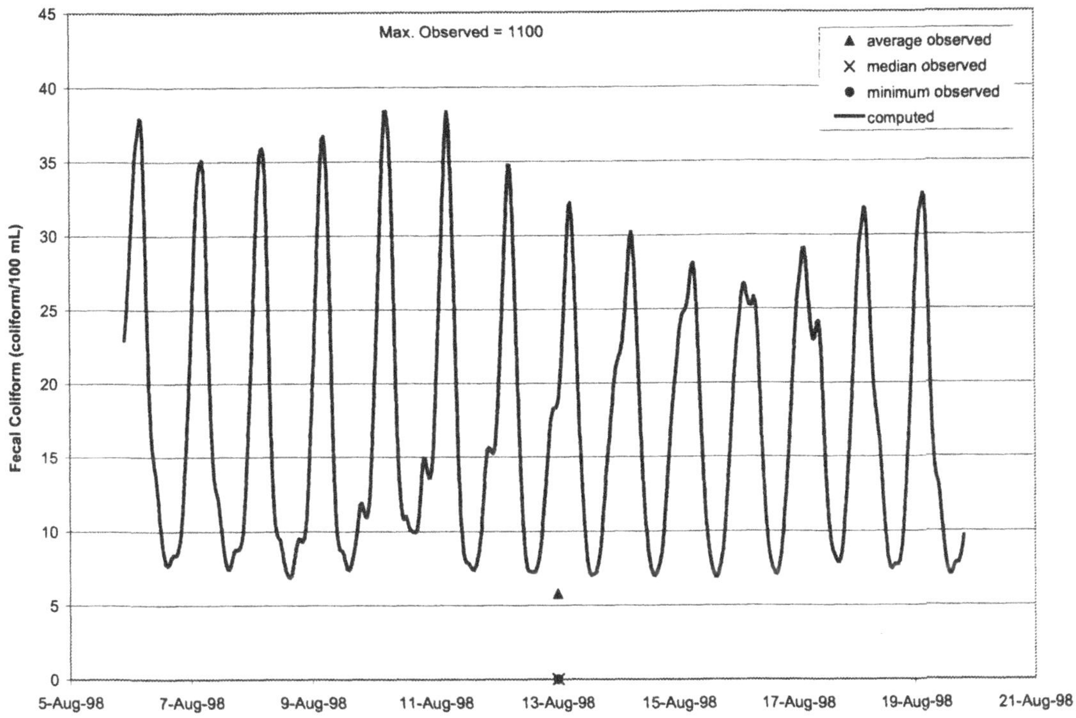


Computed and observed fecal coliform concentrations at Harbor Patrol Beach.

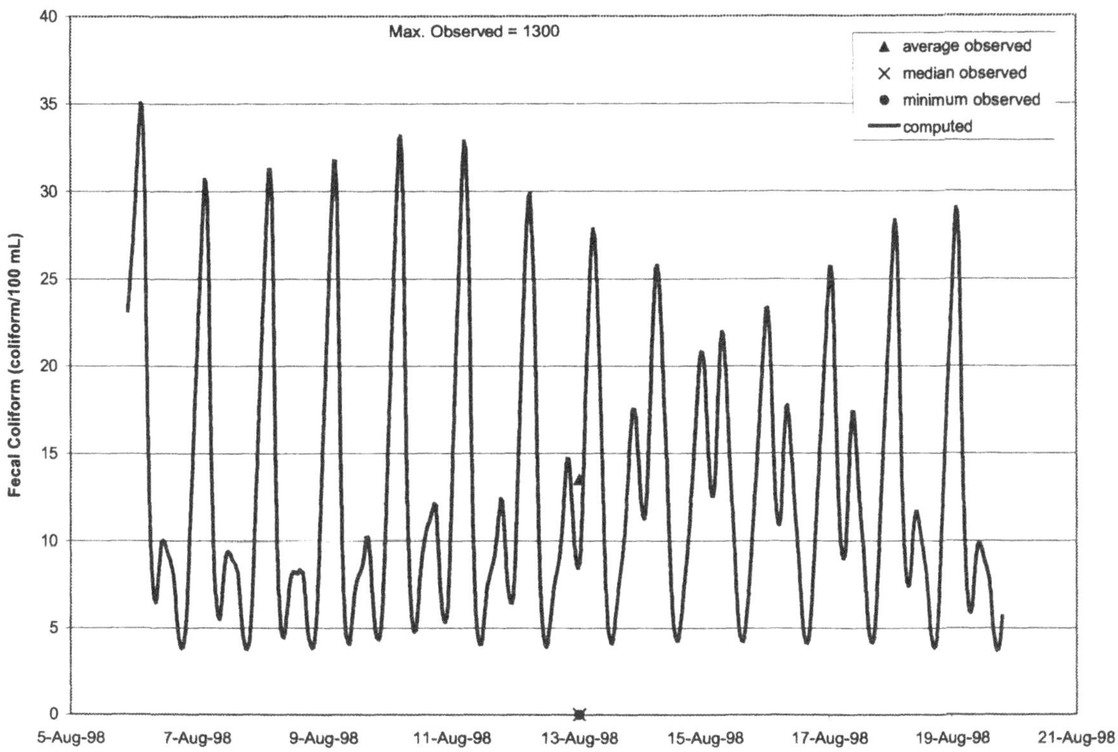

Computed and observed fecal coliform concentrations at Rocky Point Beach.

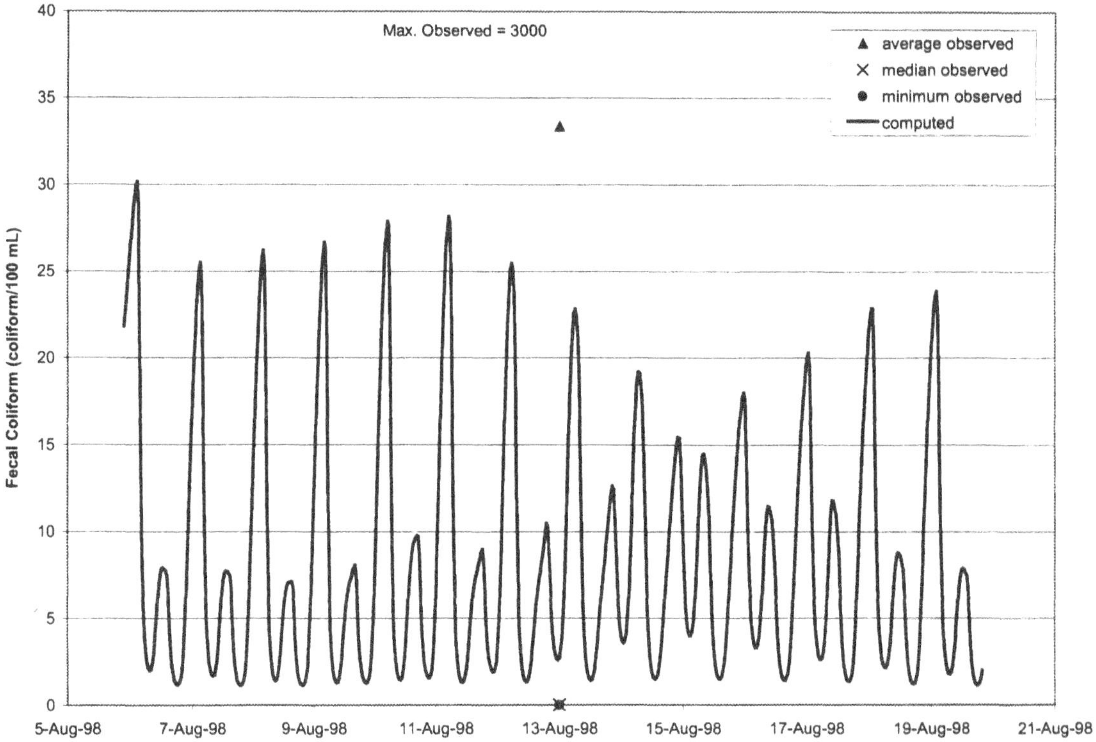


Computed and observed fecal coliform concentrations at N Street Beach

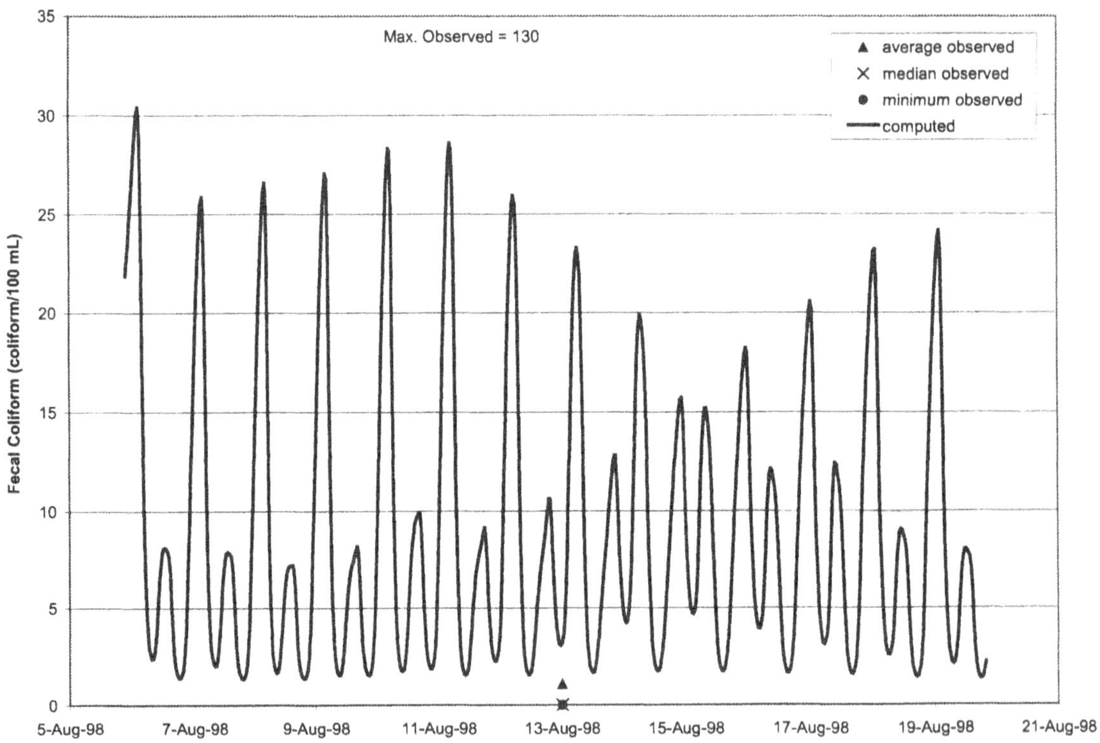

Computed and observed fecal coliform concentrations at Alvarado

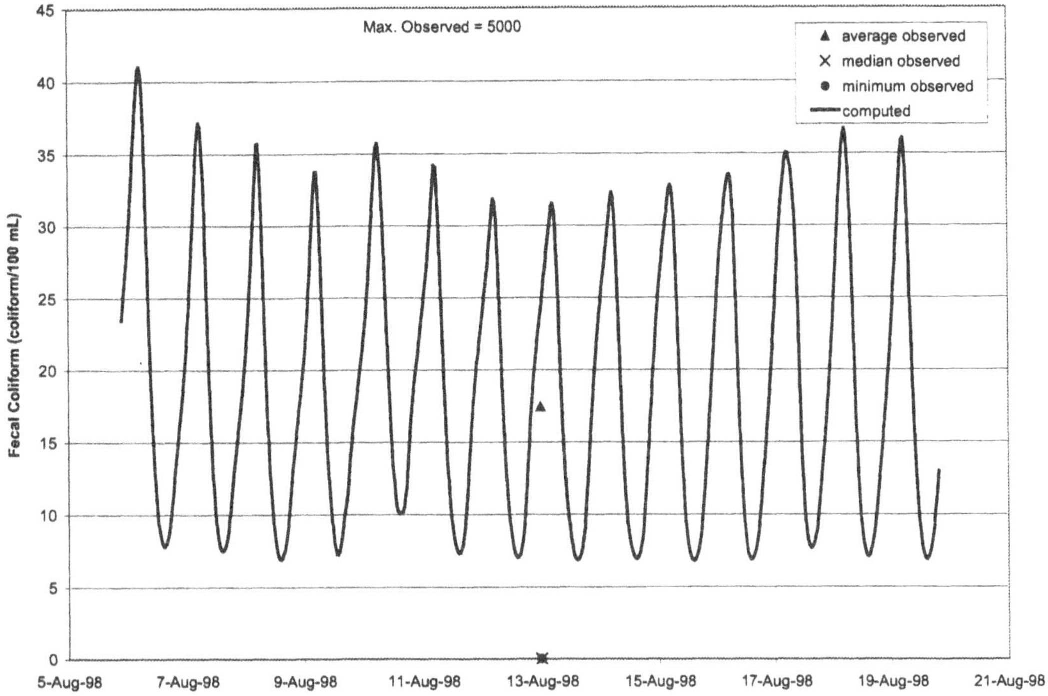


Computed and observed fecal coliform concentrations at 10th Street Beach

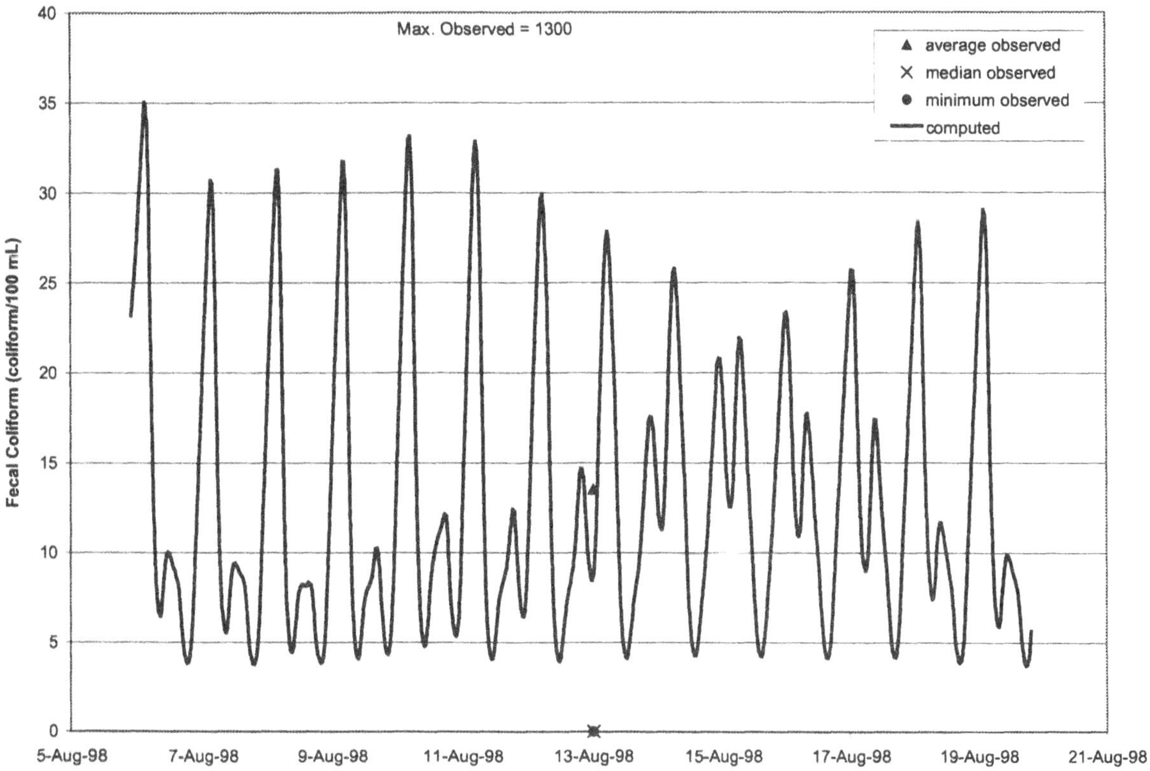

Computed and observed fecal coliform concentrations at 15th Street Beach

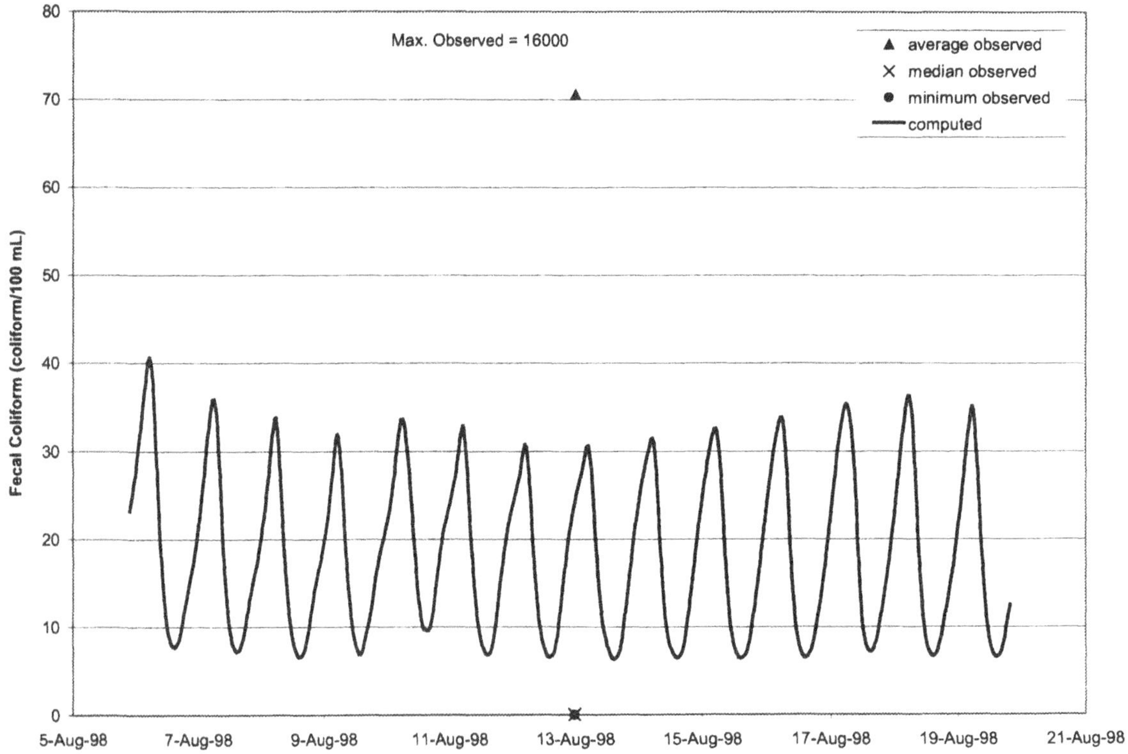


Computed and observed fecal coliform concentrations at Lido Isle Yacht Club

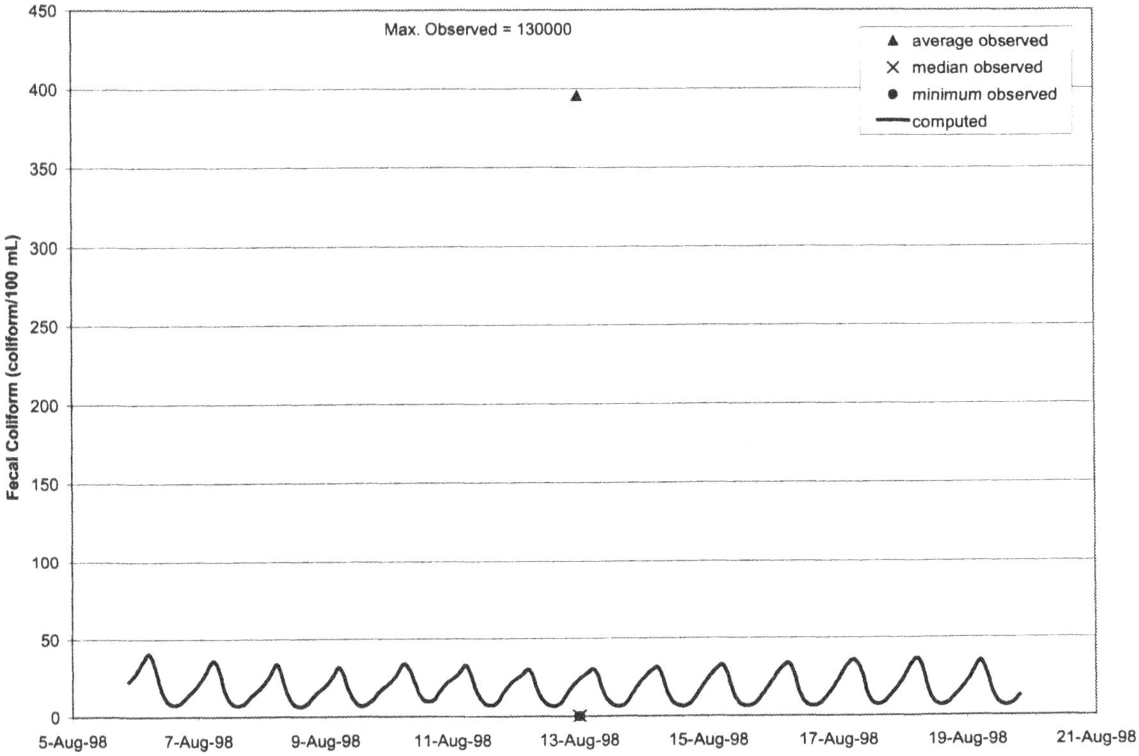

## Computed and observed fecal coliform concentrations at Via Genoa

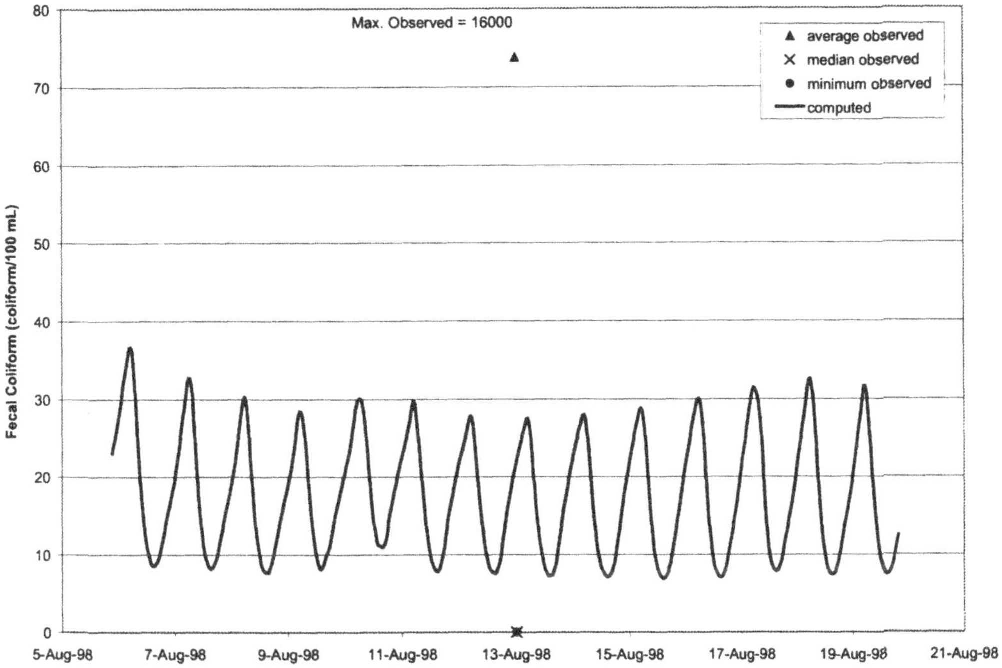


## Computed and observed fecal coliform concentrations at 19th Street Beach

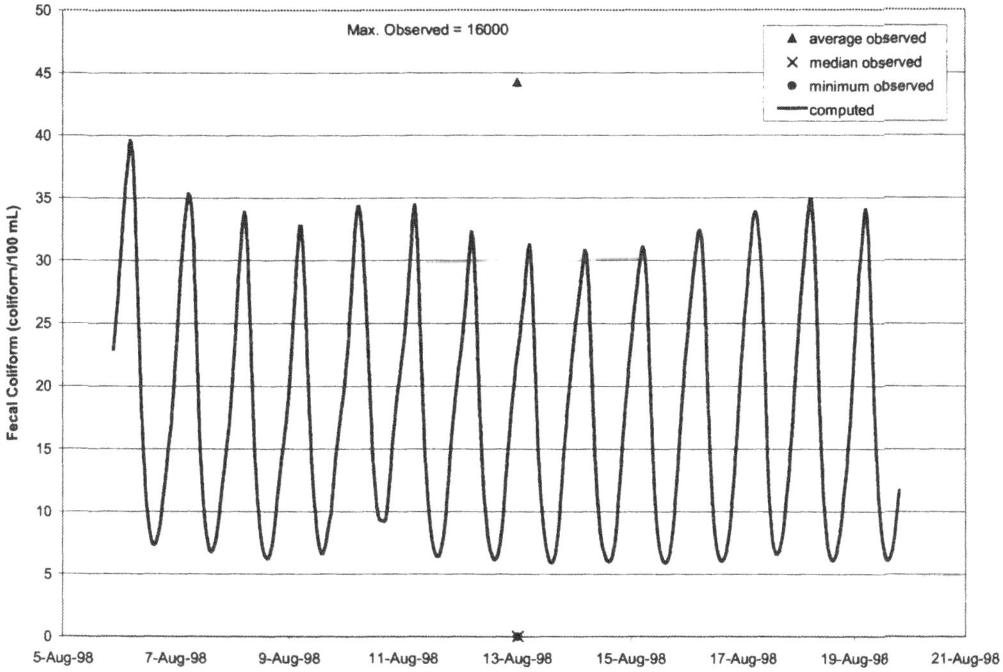

Computed and observed fecal coliform concentrations at Rhine Channel

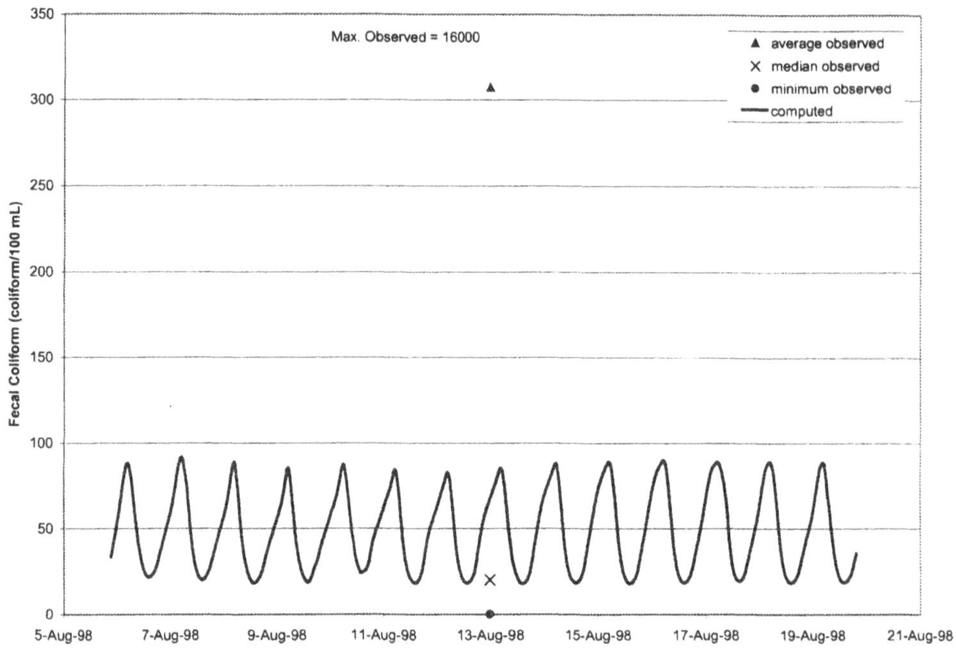


Computed and observed fecal coliform concentrations at 33rd Street Channel

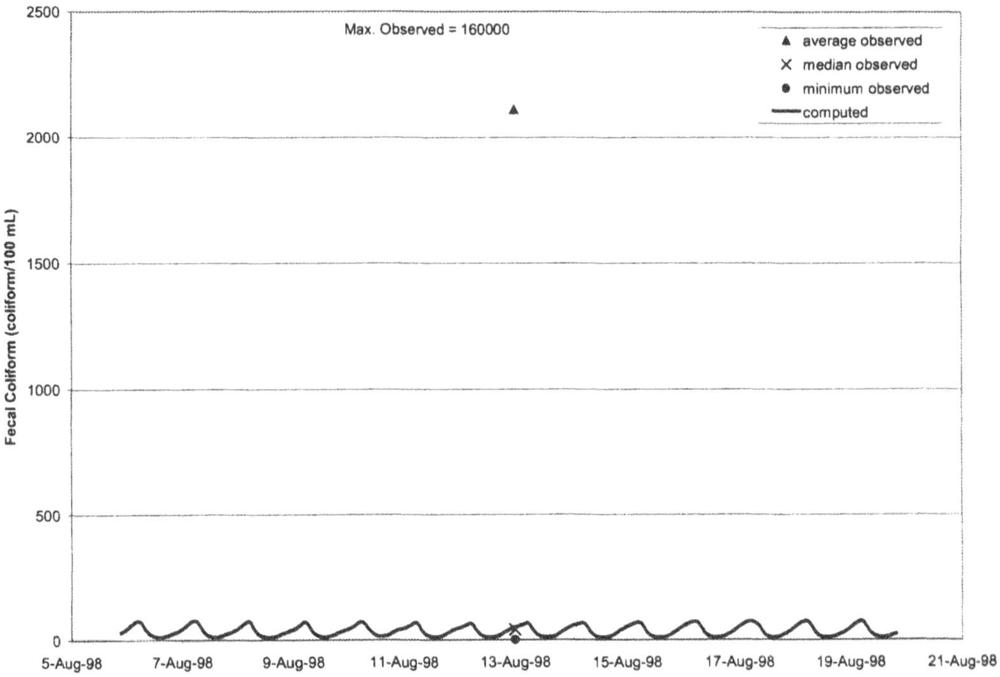


1-84339-650-5

Computed and observed fecal coliform concentrations at 38th Street Beach

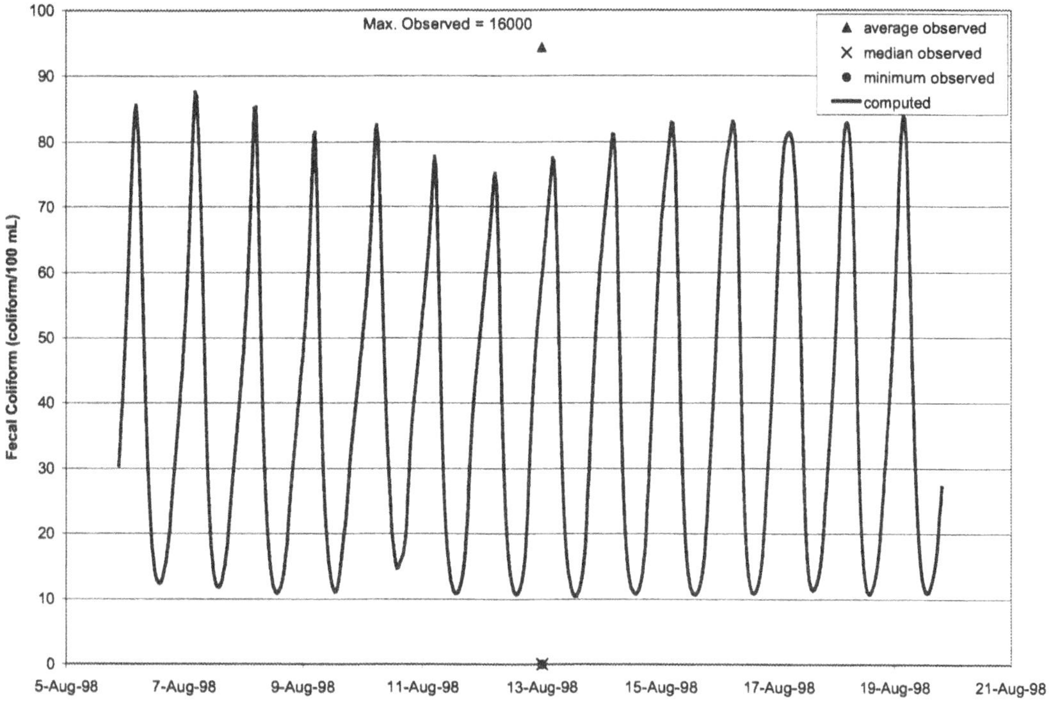


Computed and observed fecal coliform concentrations at 43rd Street Beach

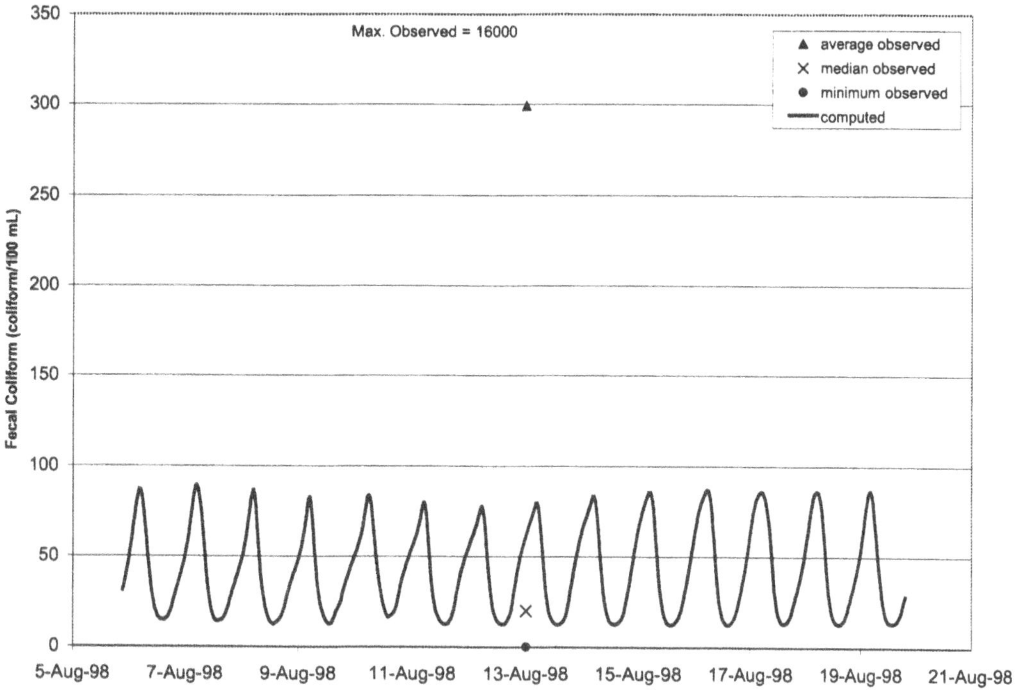

# APPENDIX B

# COMPUTED COLIPHAGE CONCENTRATIONS AT NEWPORT BAY MONITORING SITES

Computed coliphage concentrations at Ski Zone.

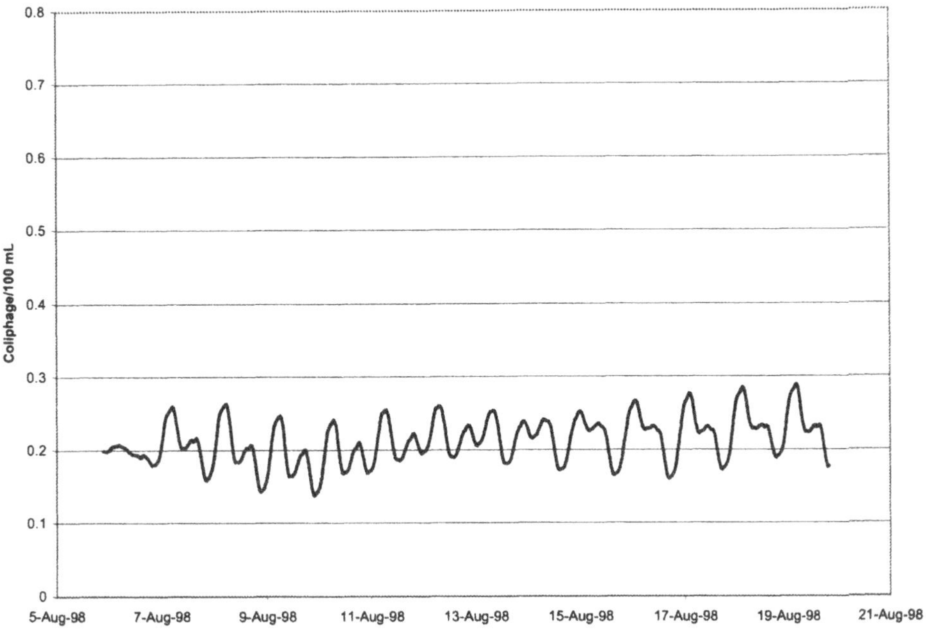


Computed coliphage concentrations at Vaughn's Launch.

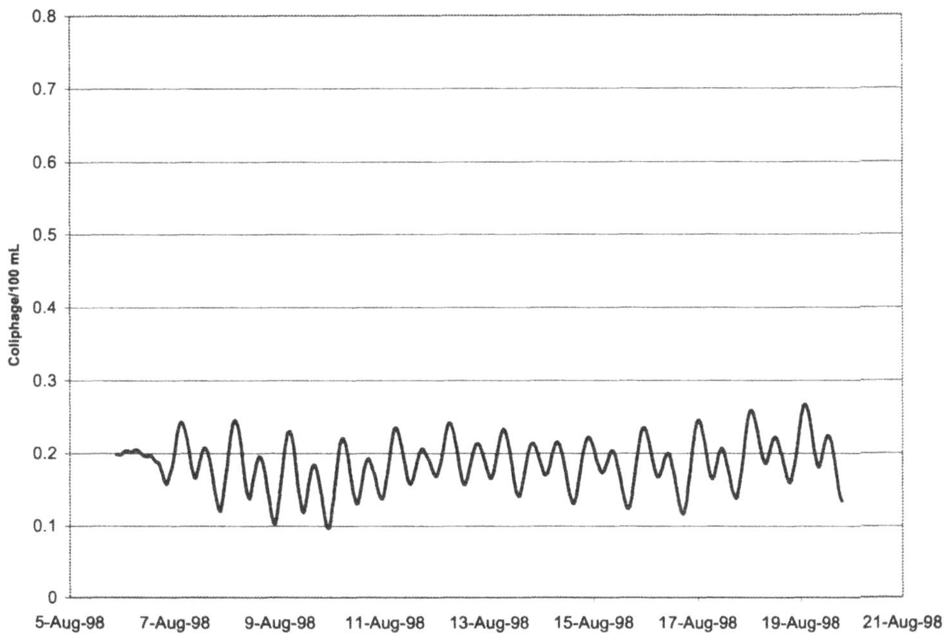

Computed coliphage concentrations at North Star Beach.

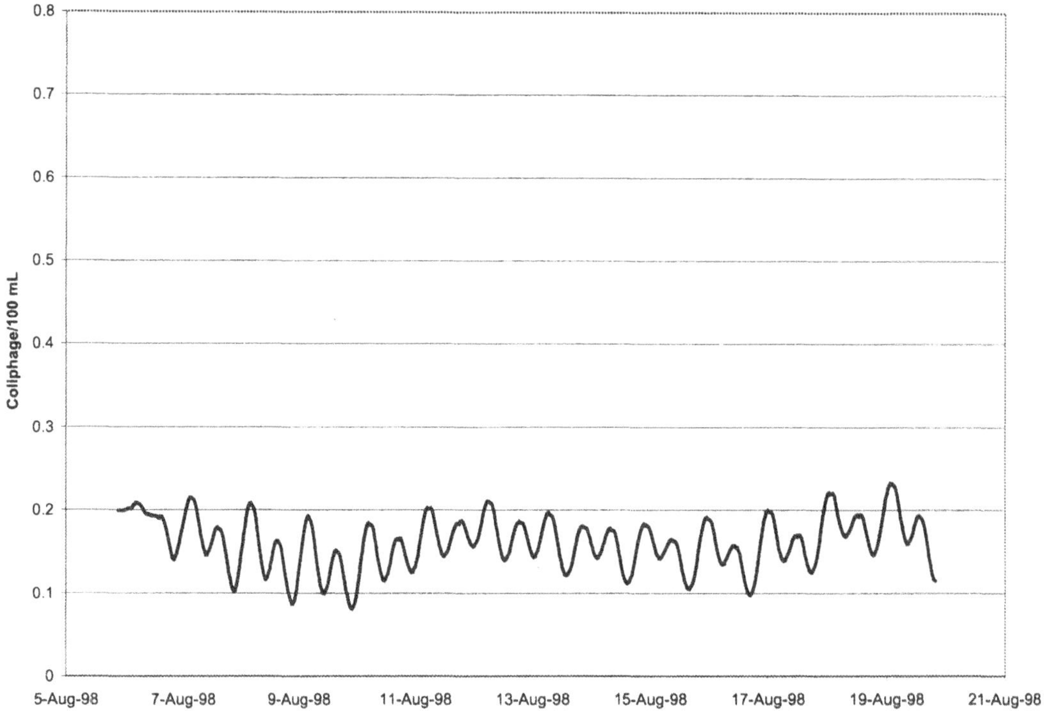


Computed coliphage concentrations at DeAnza Pier.

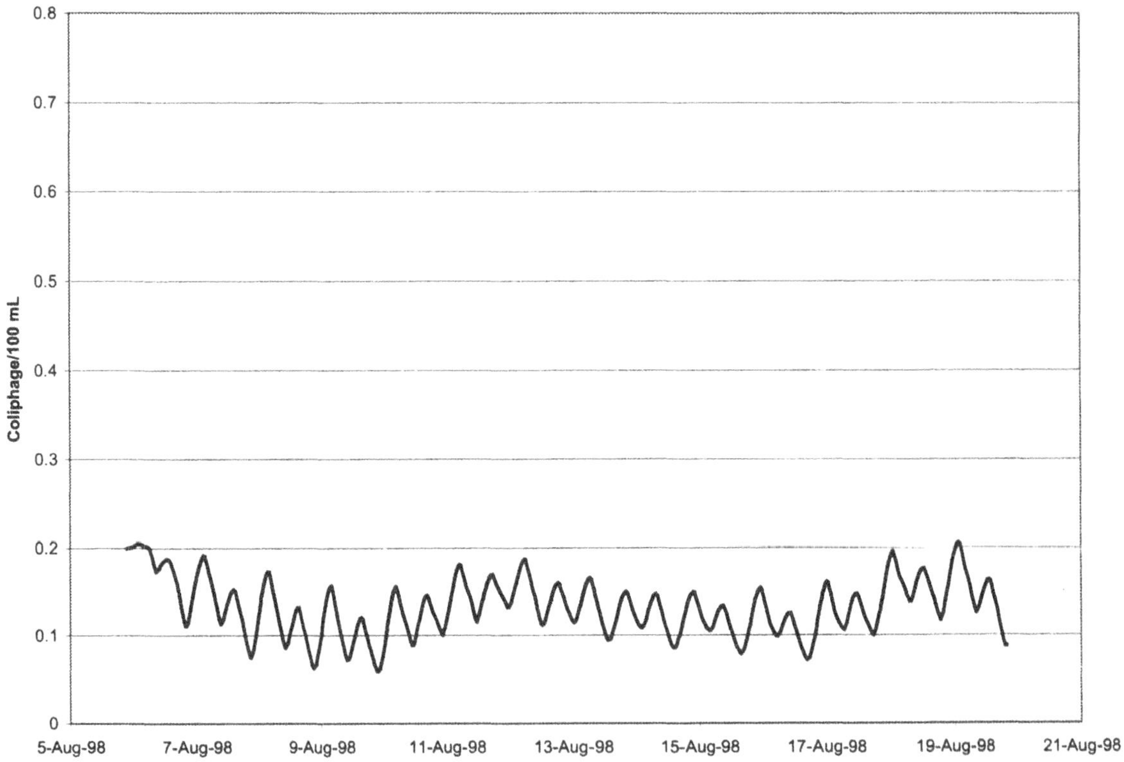

Computed coliphage concentrations at Bayshore Beach.

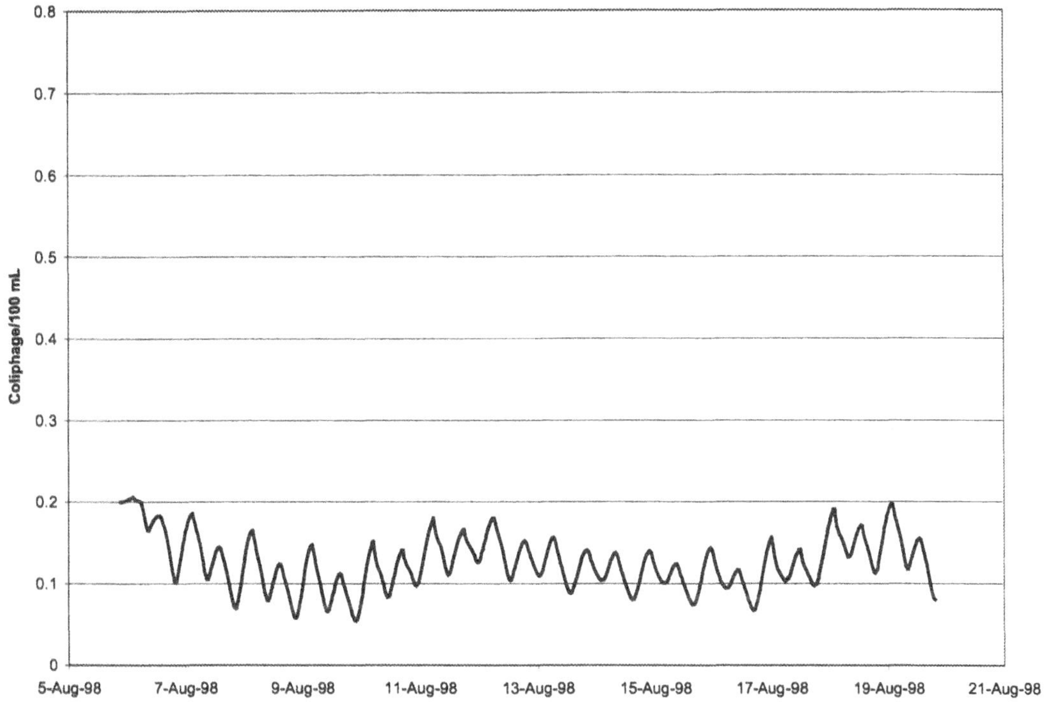


Computed coliphage concentrations at Dunes North.

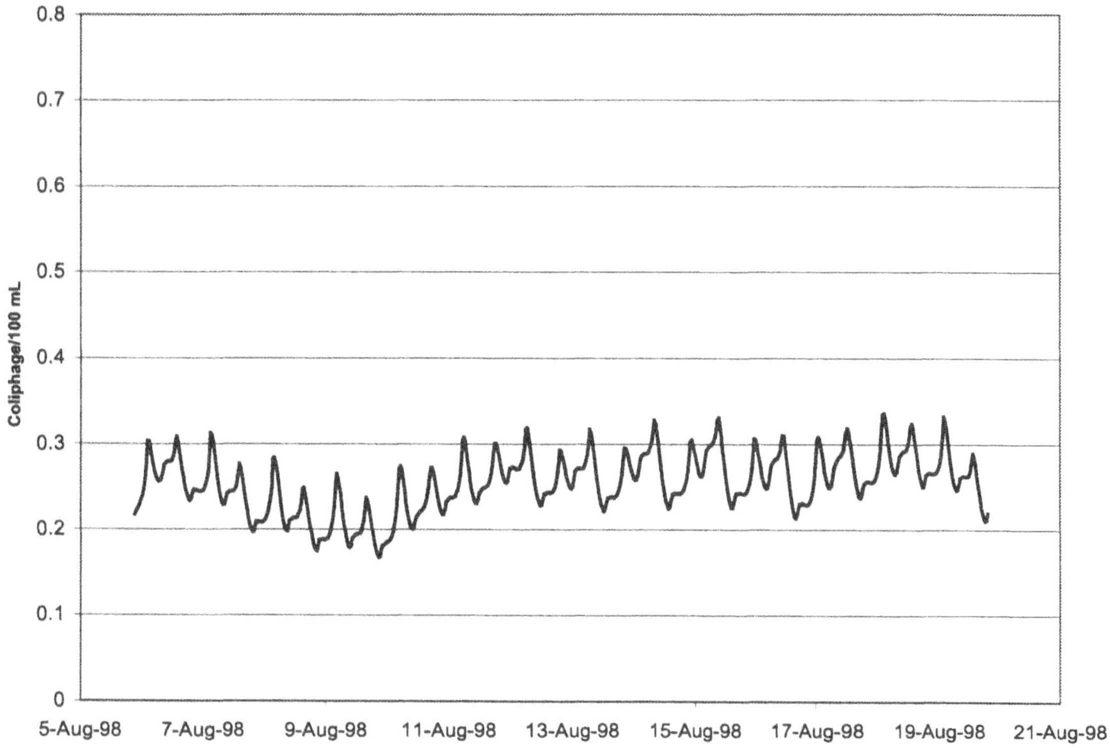

Computed coliphage concentrations at Dunes East.

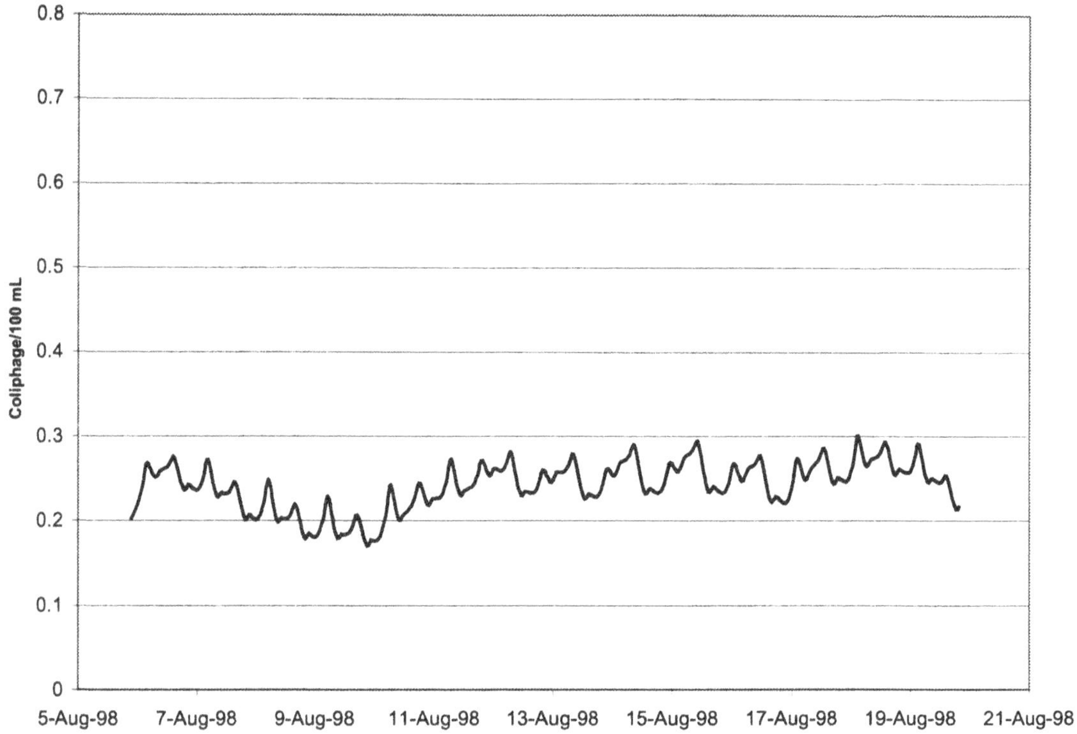


Computed coliphage concentrations at Dunes West.

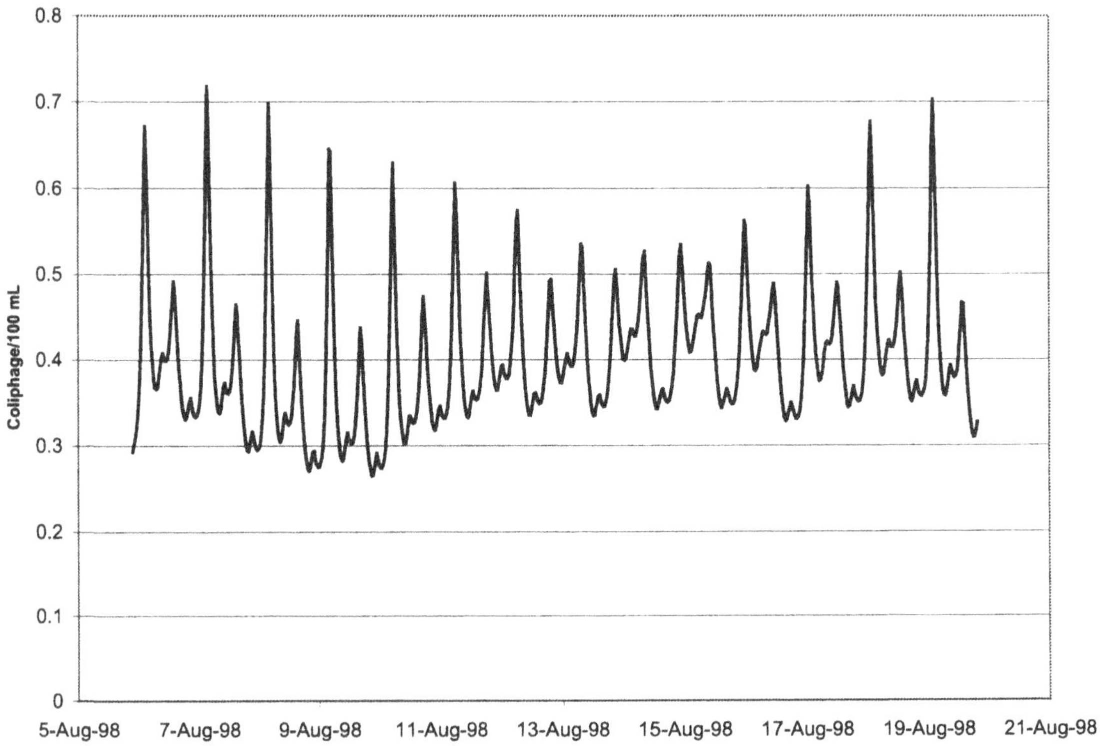

Computed coliphage concentrations at Dunes Middle.

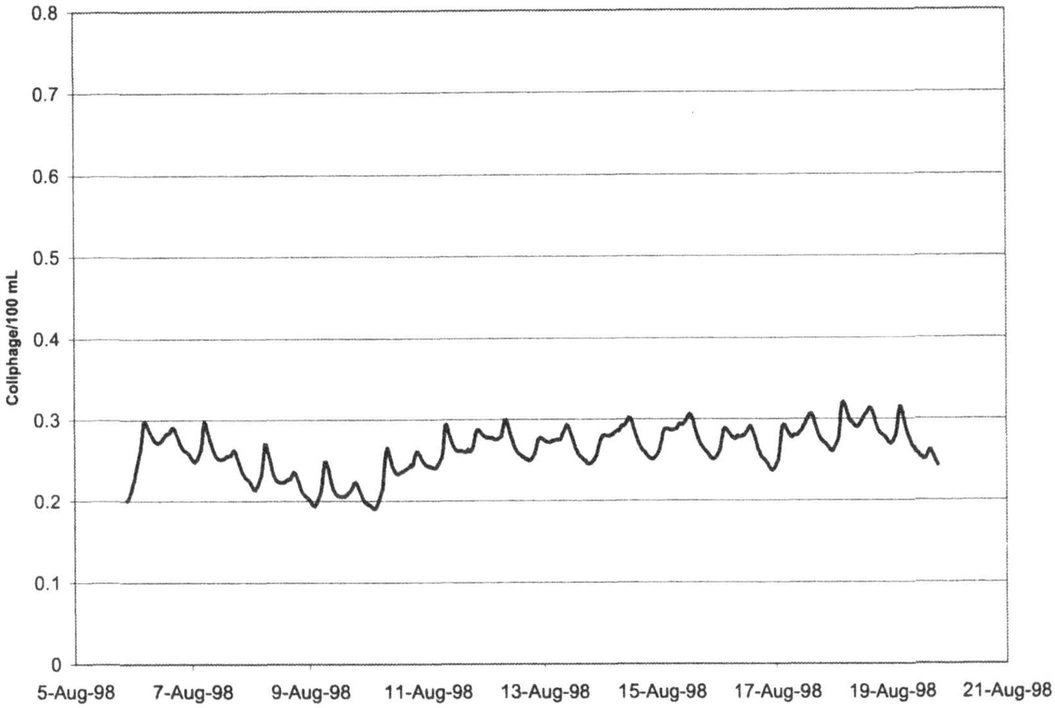


Computed coliphage concentrations at Promontory Point.

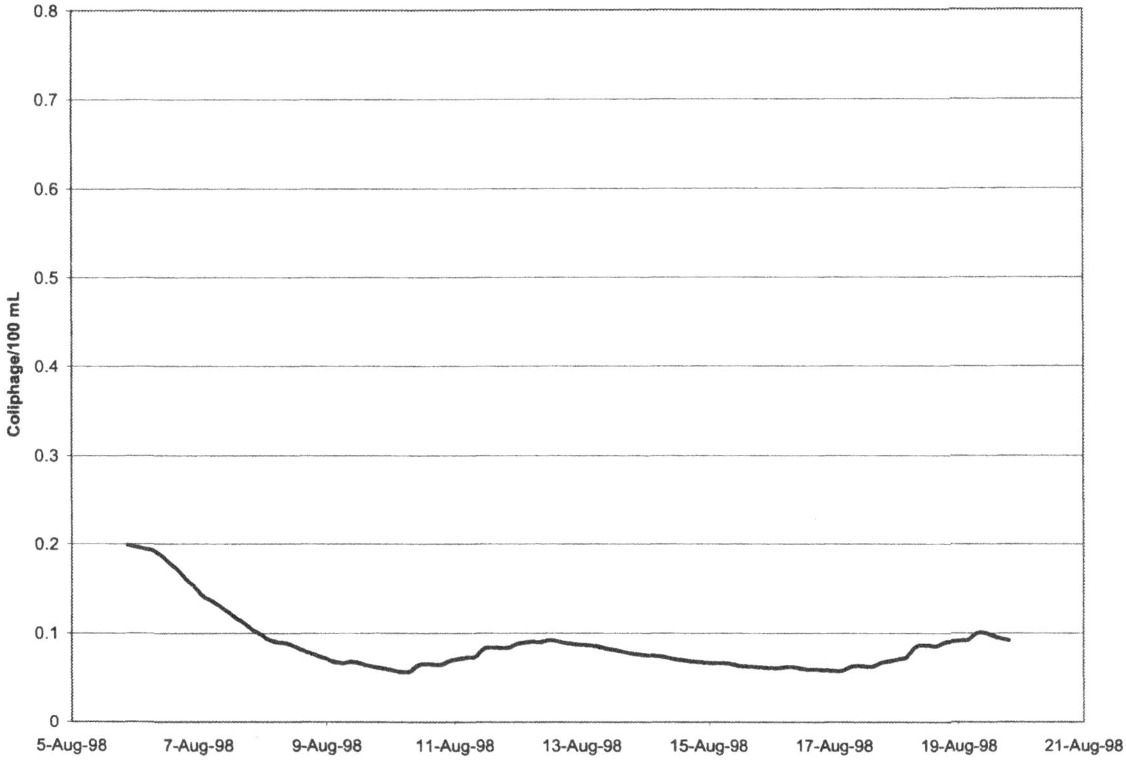

Computed coliphage concentrations at Ruby Avenue.

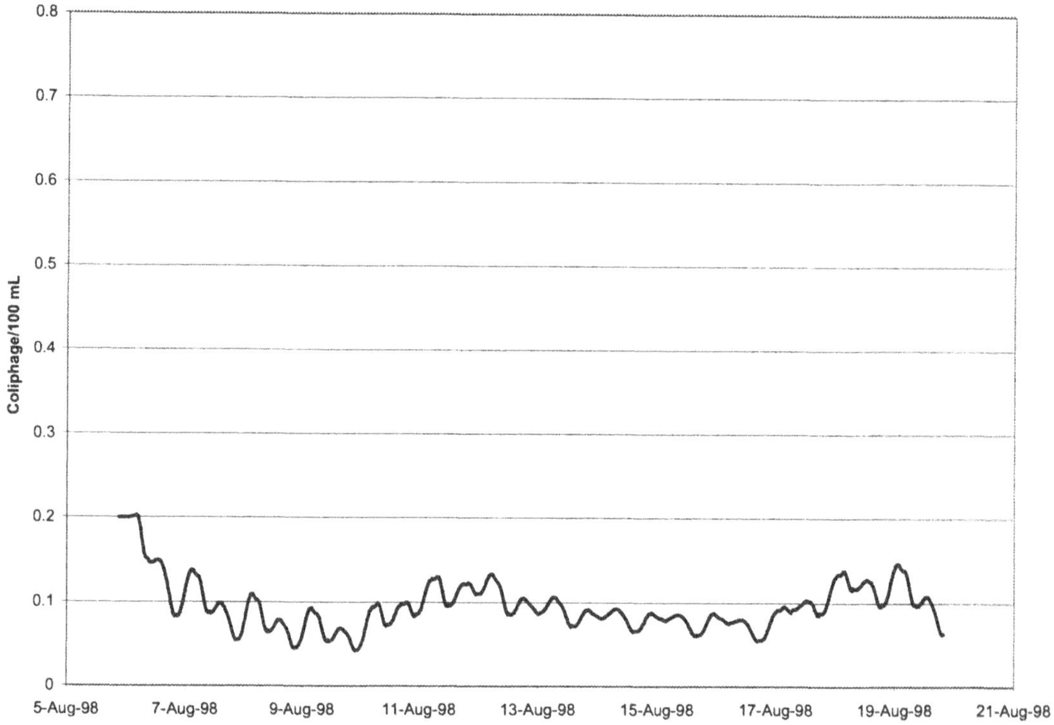


Computed coliphage concentrations at Onyx Avenue.

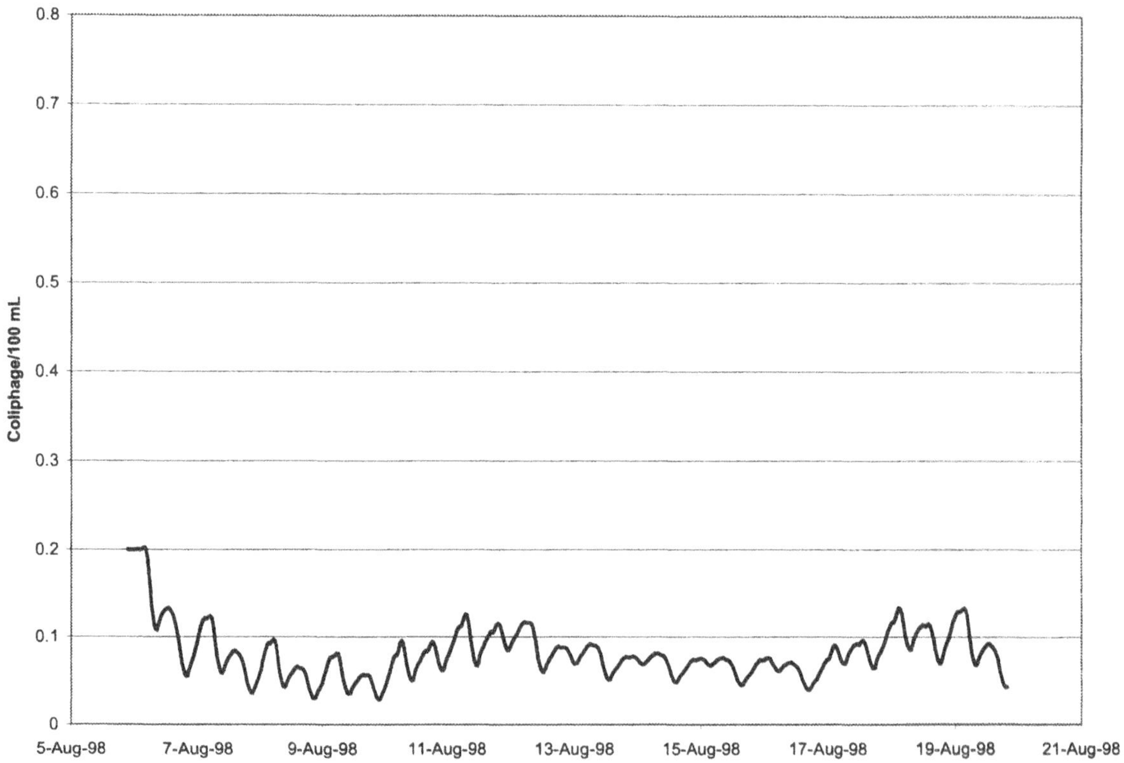

Computed coliphage concentrations at Grand Canal

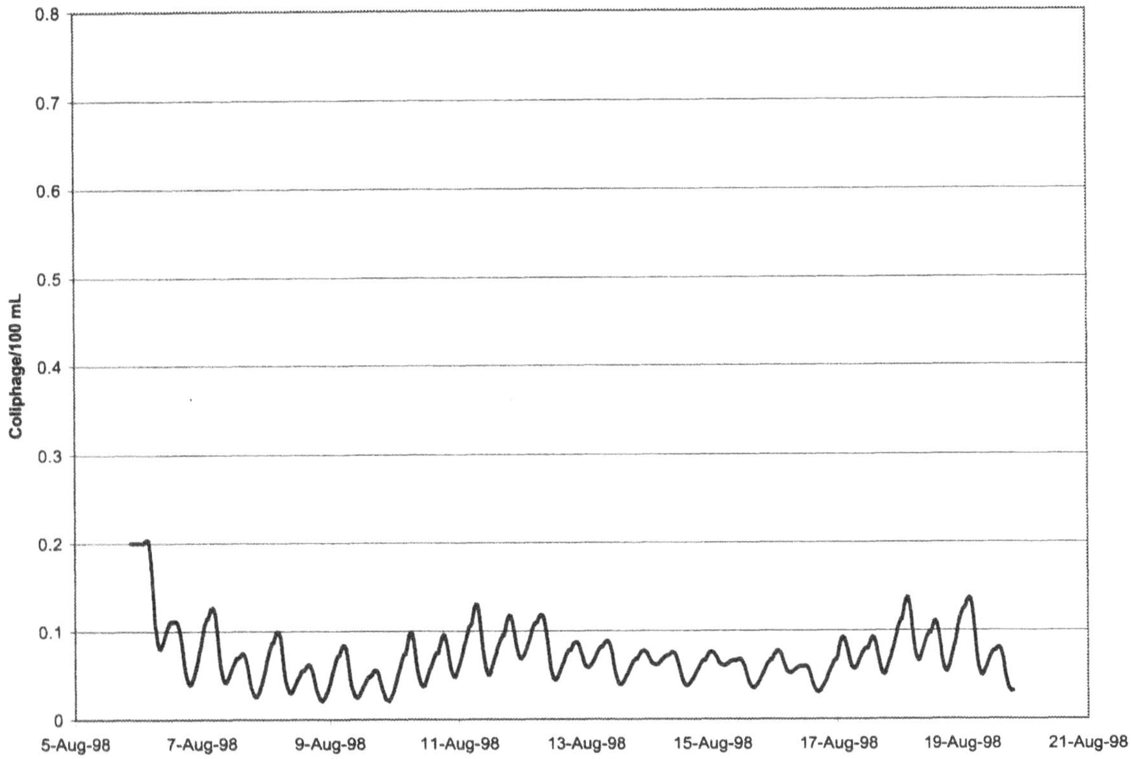


Computed coliphage concentrations at Park Avenue.

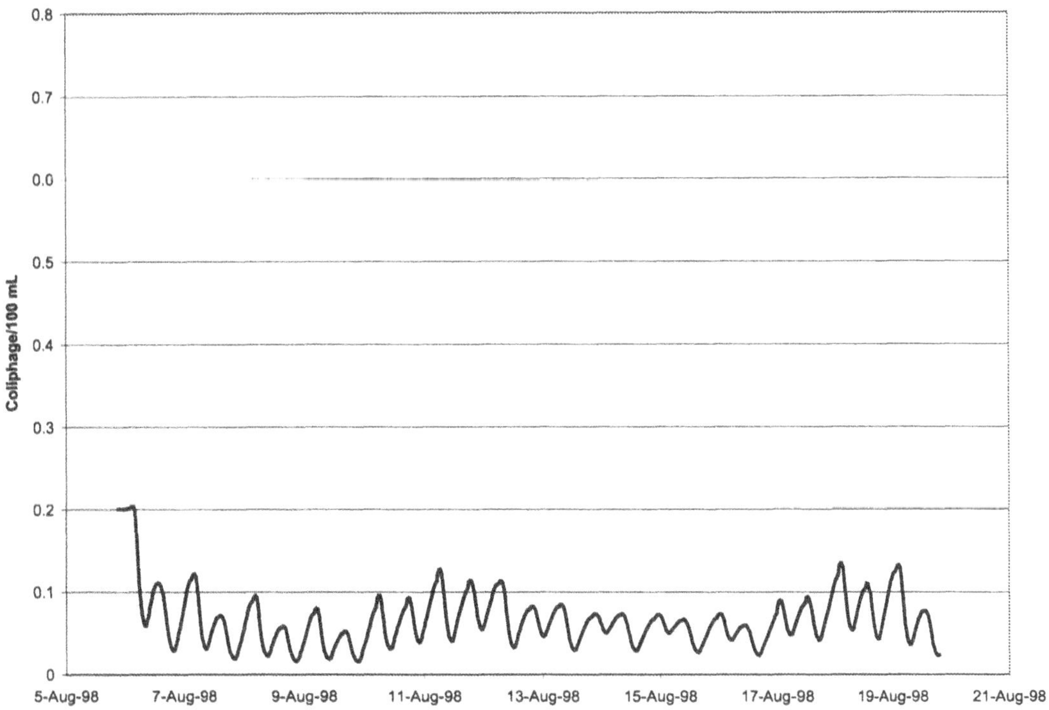

Computed coliphage concentrations at Abalone Avenue

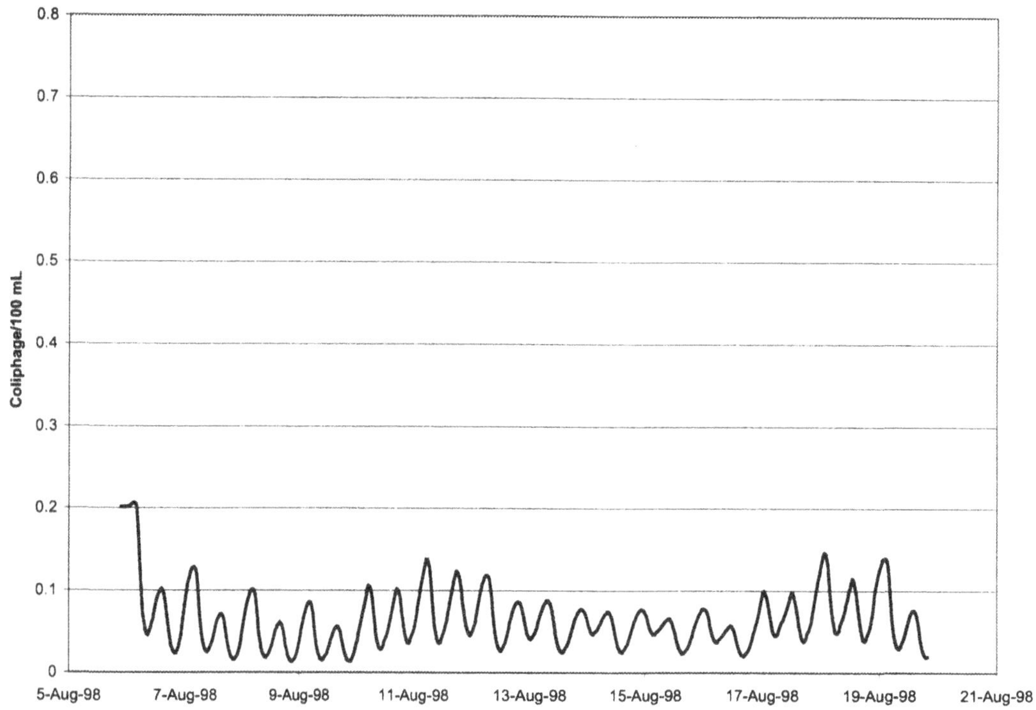


Computed coliphage concentrations at Sapphire Avenue

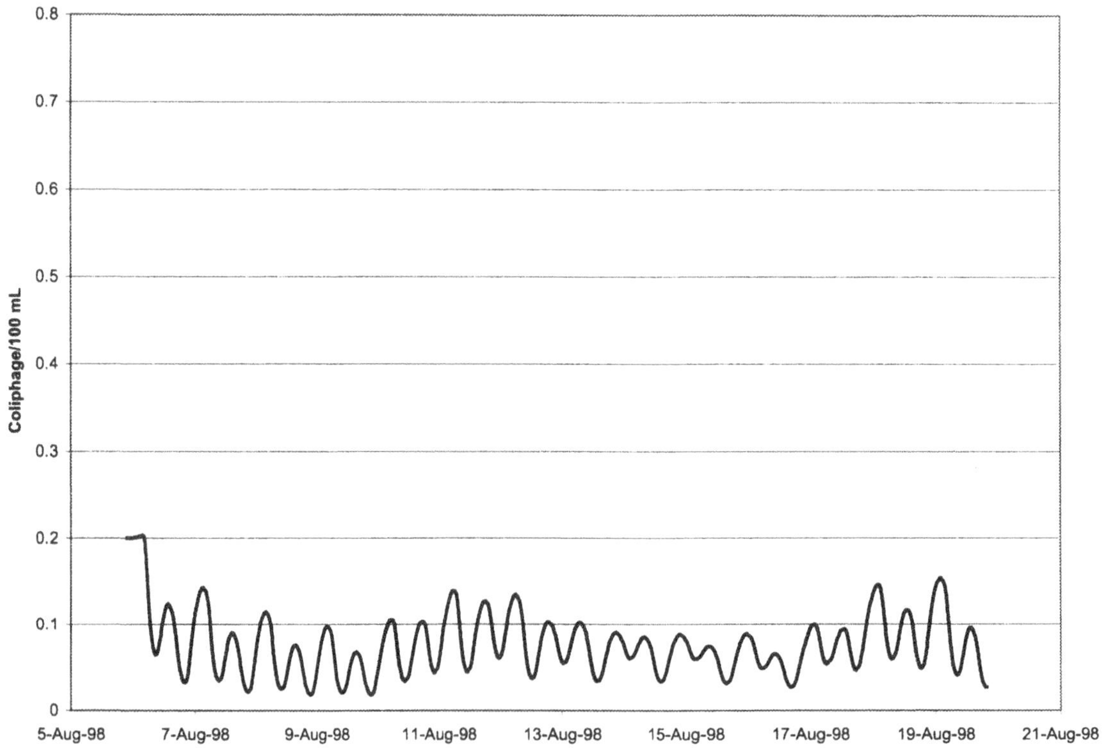

Computed coliphage concentrations at Garnet Avenue.

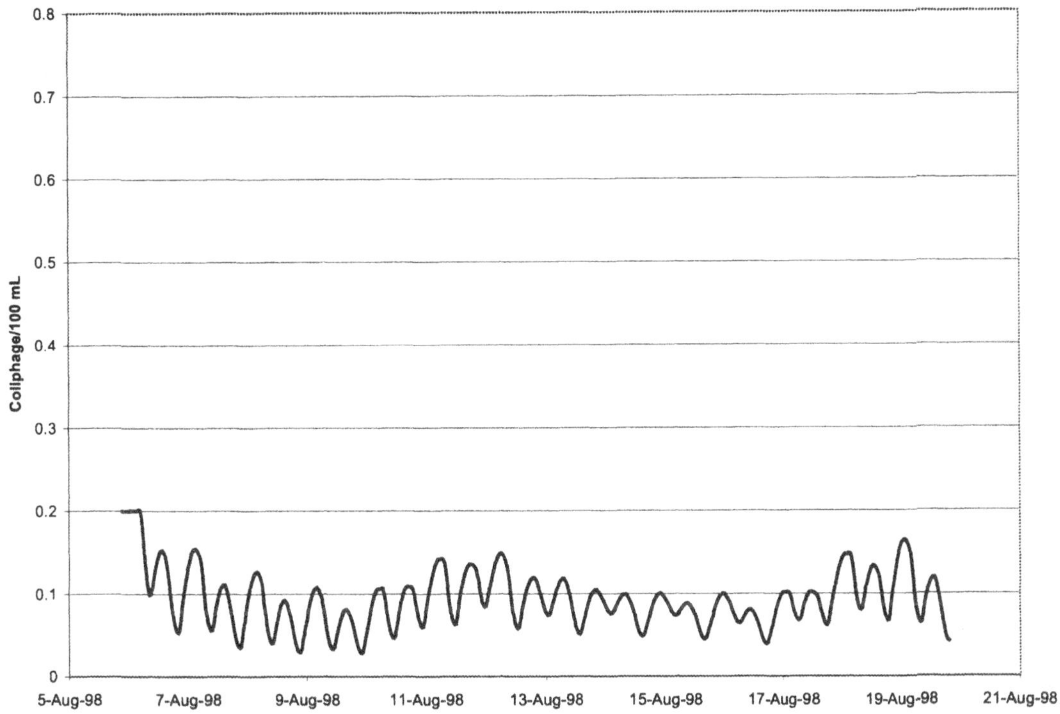


Computed coliphage concentrations at Harbor Patrol Beach.

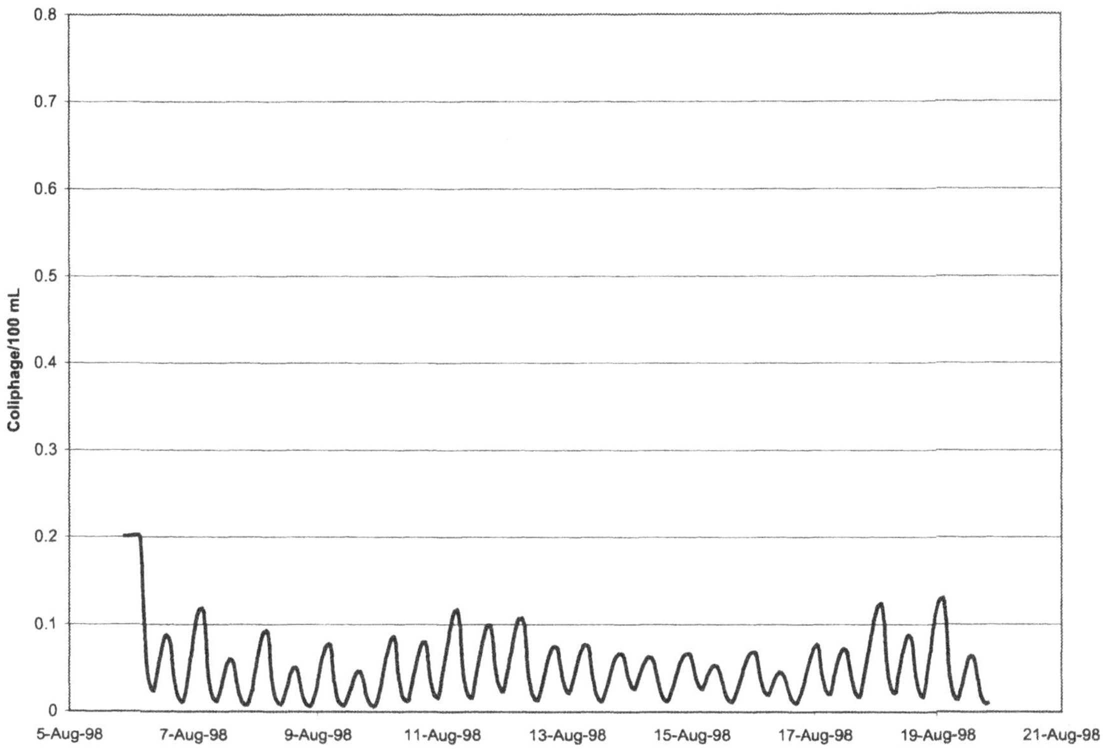

Computed coliphage concentrations at Rocky Point Beach.

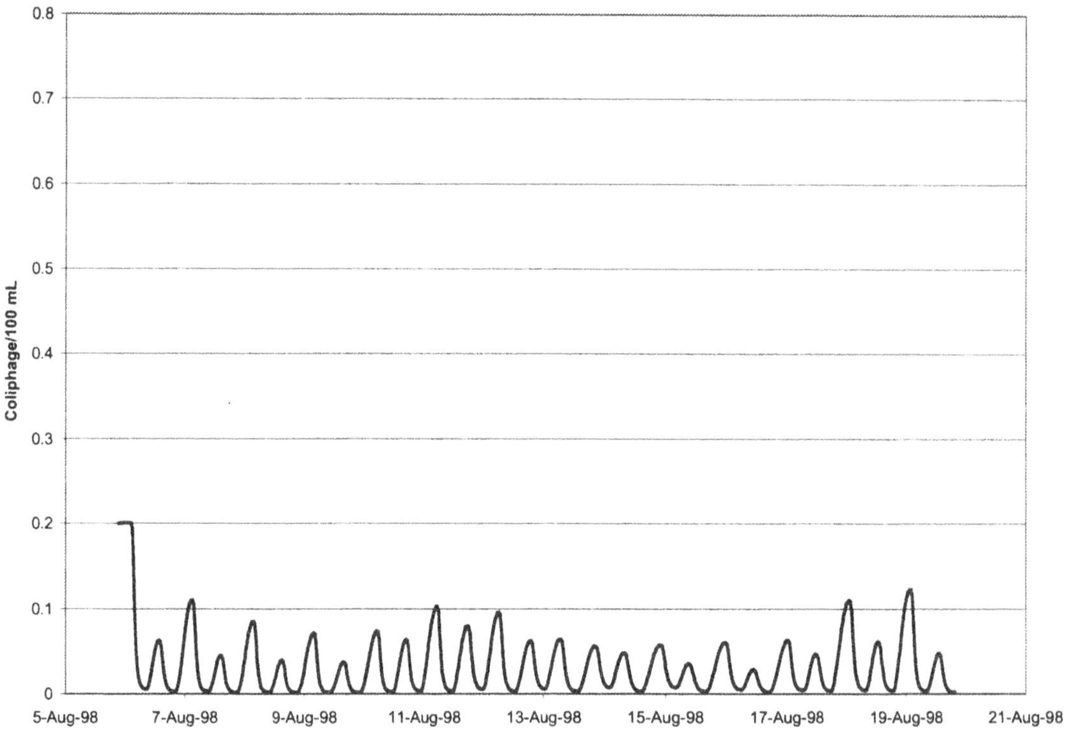


Computed coliphage concentrations at N Street Beach.

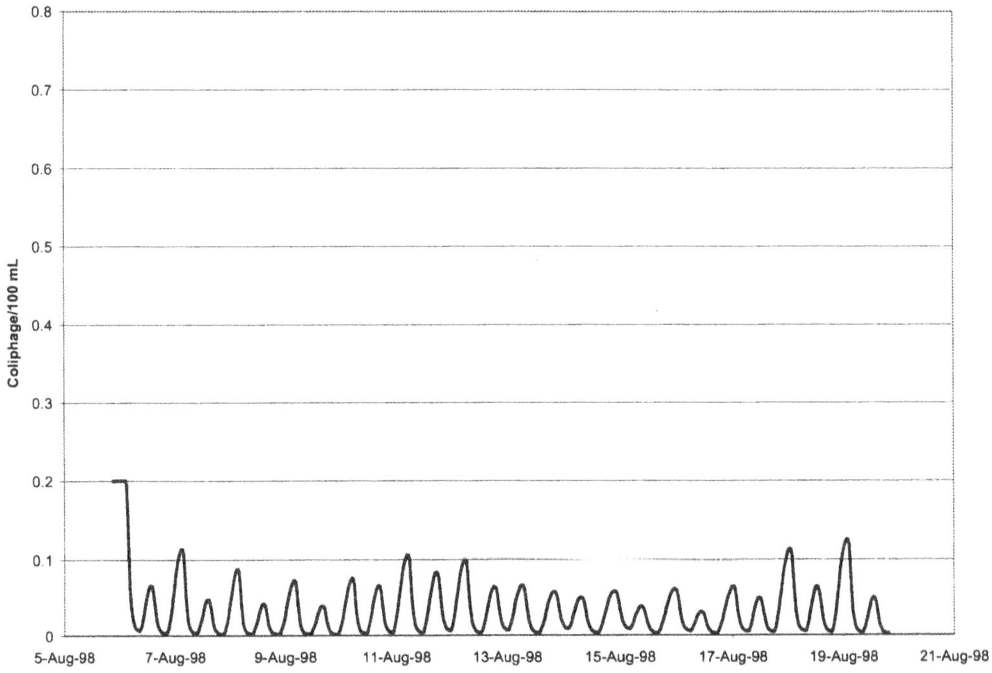

Computed coliphage concentrations at Alvarado.

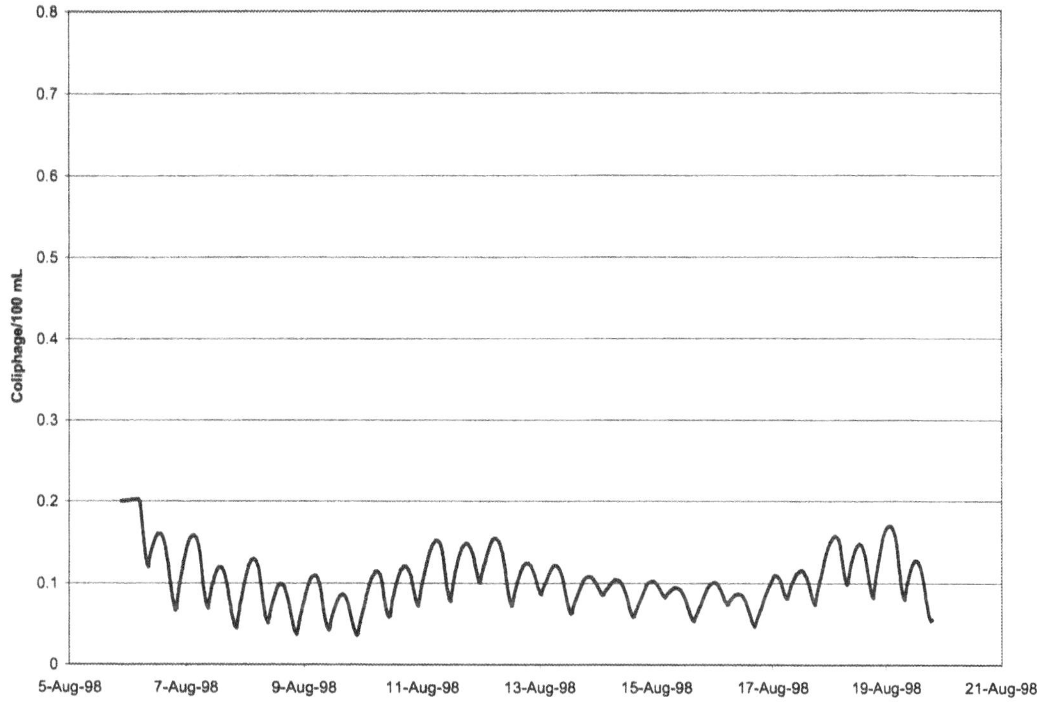


Computed coliphage concentrations at 10th Street Beach.

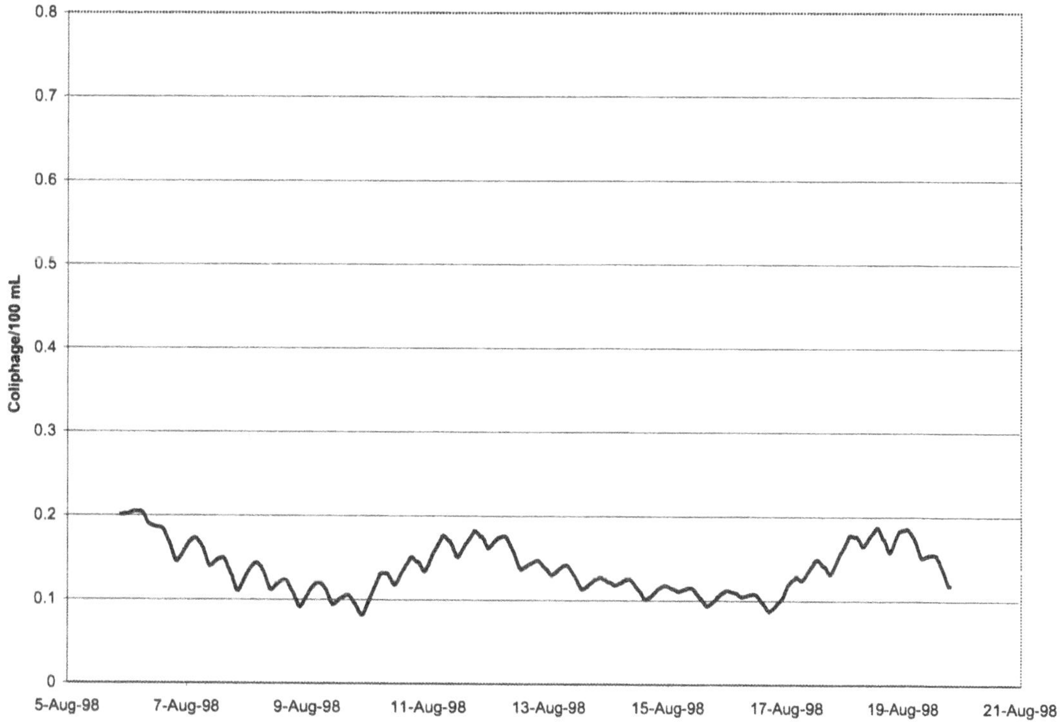

## Computed coliphage concentrations at 15th Street Beach.

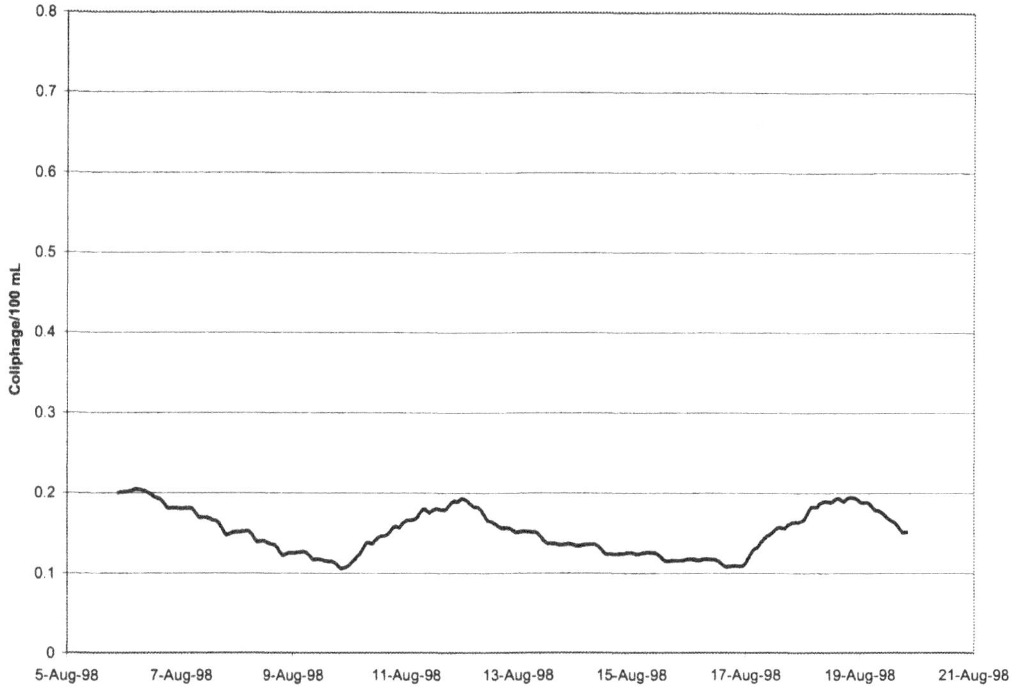


## Computed coliphage concentrations at Lido Isle Yacht Club.

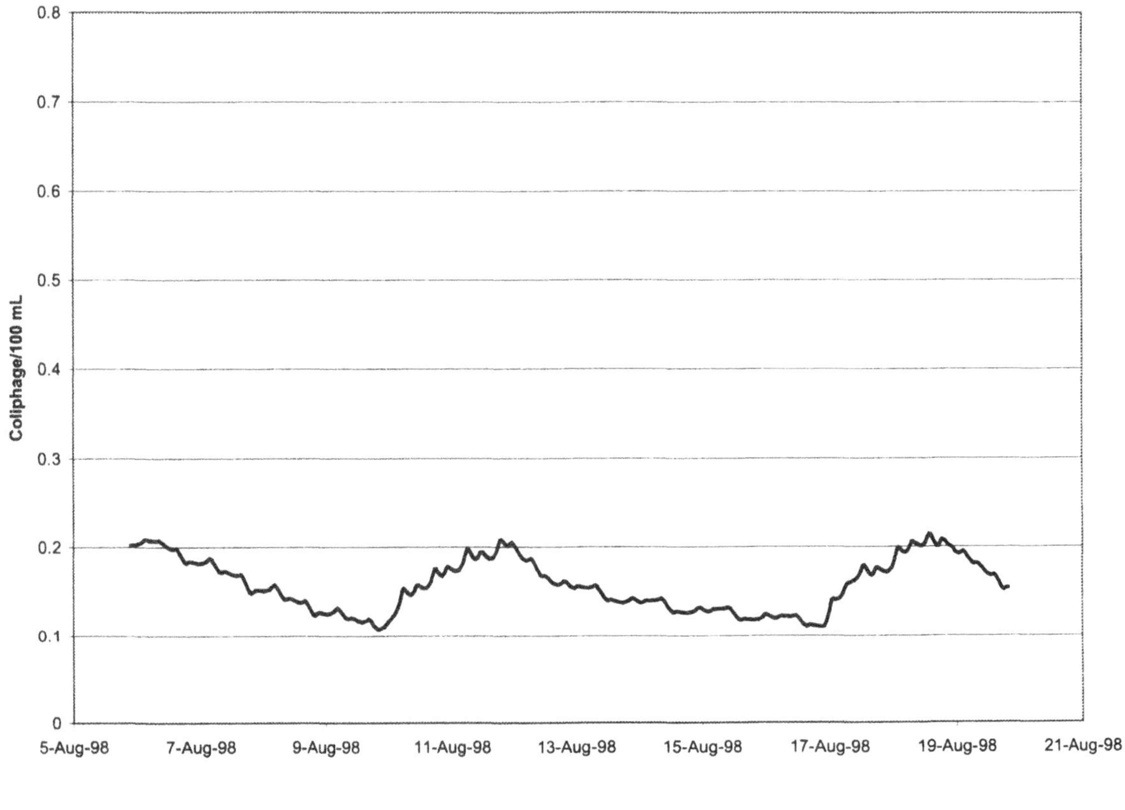

Computed coliphage concentrations at Via Genoa.

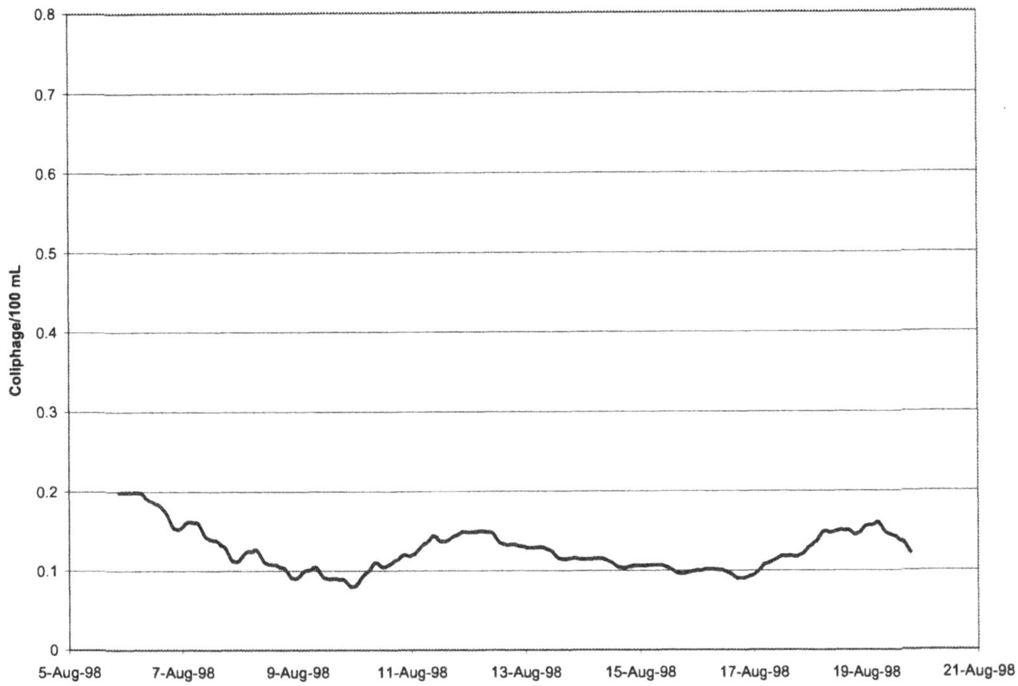


Computed coliphage concentrations at 19th Street Beach.

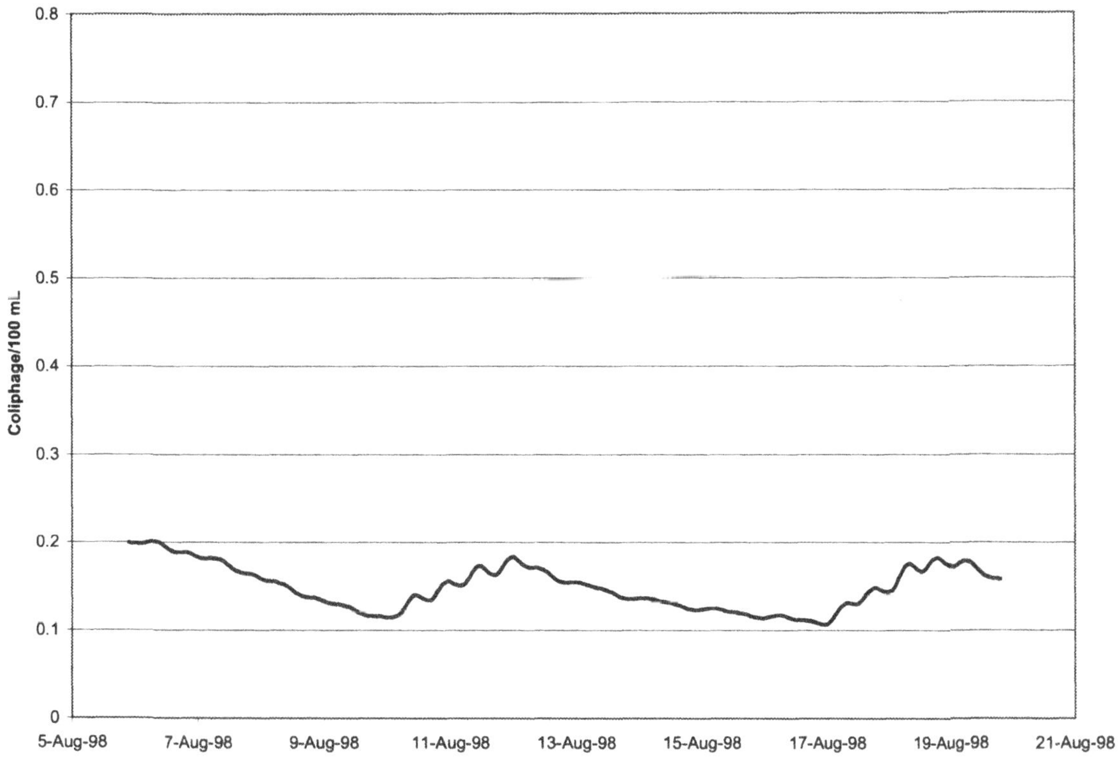

Computed coliphage concentrations at Rhine Channel.

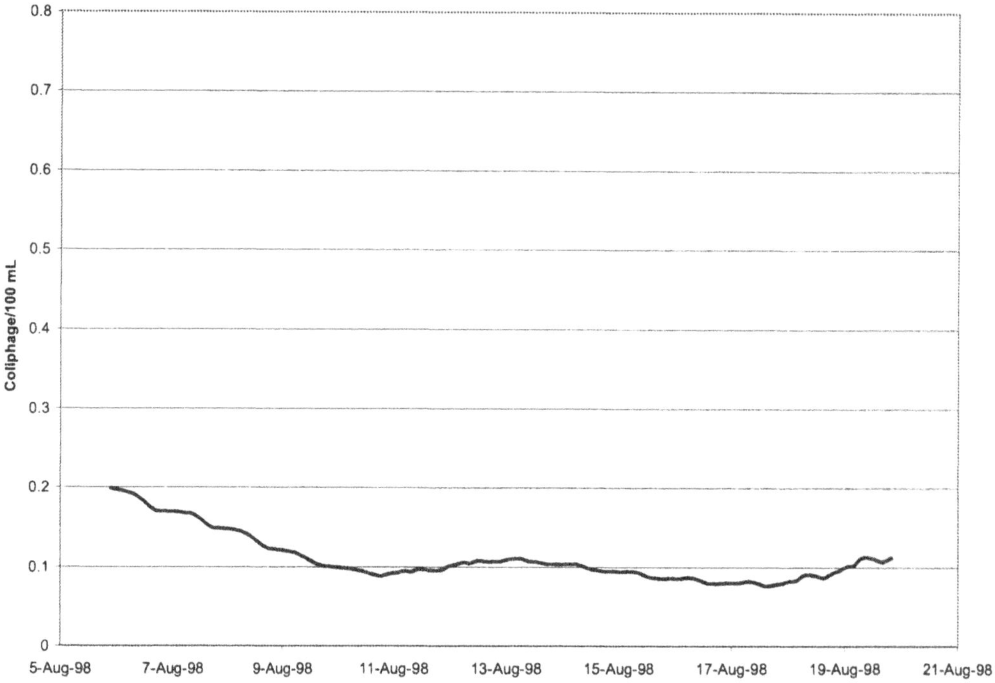


Computed coliphage concentrations at 33rd Street Channel.

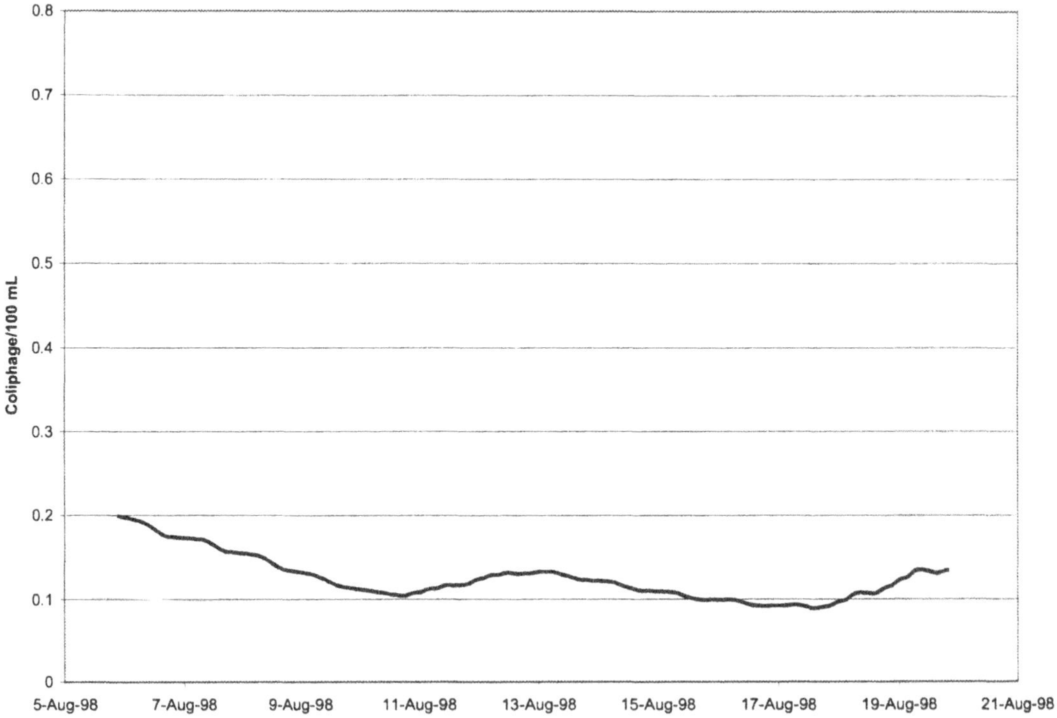

Computed coliphage concentrations at 38th Street Beach.

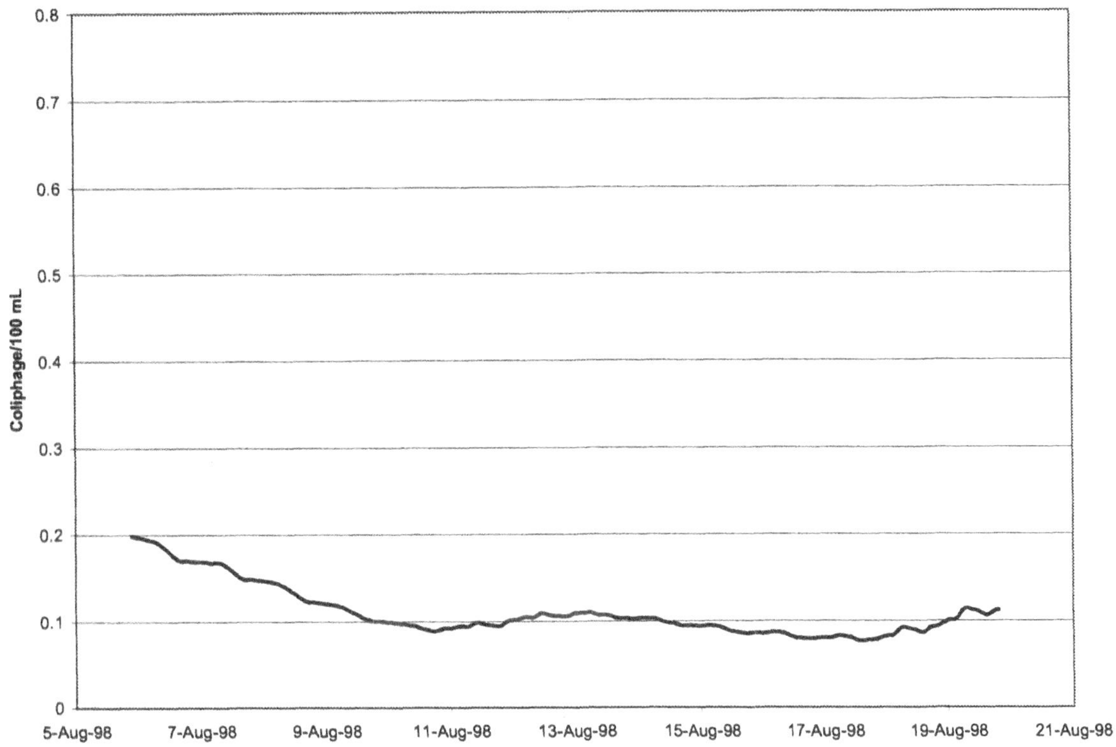


Computed coliphage concentrations at 43rd Street Beach.

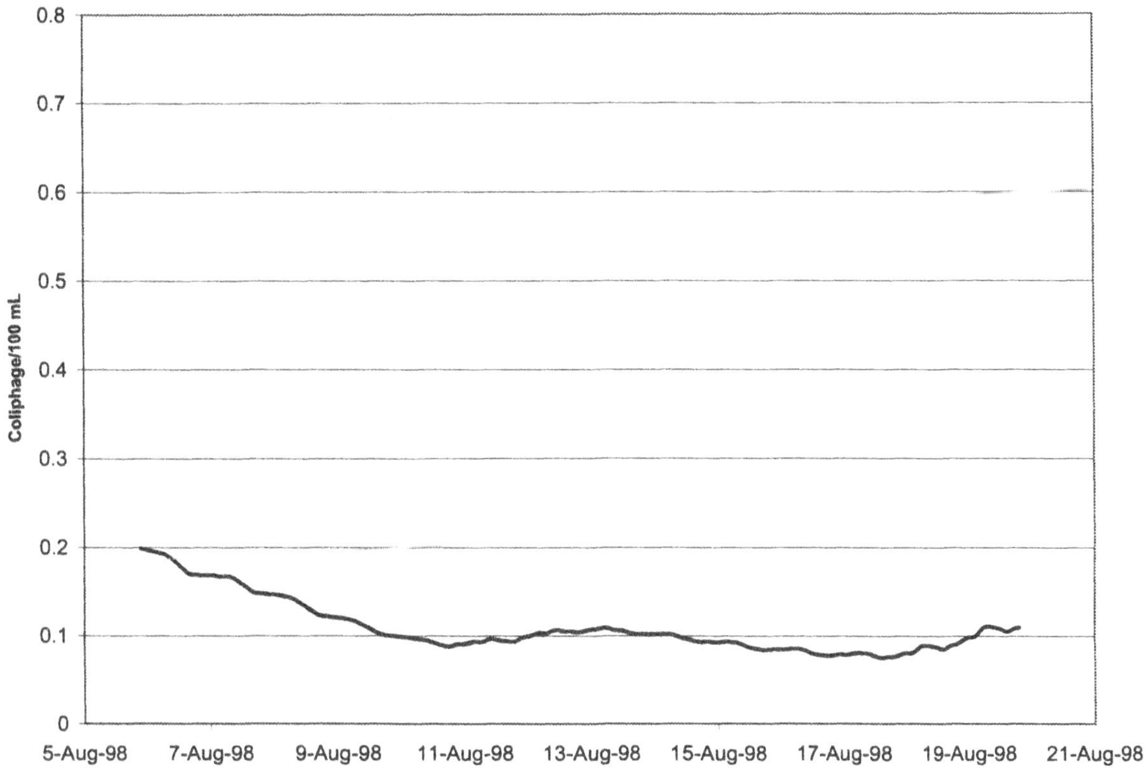

# REFERENCES


APHA, AWWA, and WPCF. (1995). "Standard Methods for the Examination of Water and Wastewater. (19th edition)". Am. Public Health Assoc., 1015 Fifteenth St. NW, Washington, DC 20005.

Armon, R., and Kott, Y. (1996). "Bacteriophages as Indicators of Pollution". *Critical Reviews in Environmental Science and Technology*, Vol. 26, No. 4, pp. 299-335.

AWWARF. (1999). "Using Microbial Indicators in Optimization of Treatment Plant Performance". E. Nieminski, Presented at the International Symposium on Waterborne Pathogens, AWWA, Milwaukee, WI.

Baldini, I., Scandelari, E., and Cacciabue, E. (1978). "Research of the RNA-coliphages, in comparison with usual microbial and chemical indexes of biological pollution for hygienic and sanitary evaluations of coast waters". *Ann Sclavo*, Vol. 20, No. 4, pp. 588-598.

Black, S. A., Graveland, D. N., Nicholaichuk, W., Smith, D. W., and Tobin, R. S. (1984). "Manual for Land Application of Treated Municipal Wastewater and Sludge". Environmental Protection Service, Ottawa (Ontario). Environmental Protection Programs Directorate. Manual EPS 6-EP-84-1, March, 1984. p 216, 29 Fig, 70 Tab, 142 Ref, 4 Append., Journal Announcement: SWRA1808.

Borrego, J. J., Morinigo, M. A., De Vincente, A., C, Ornax, R., and Romero, P. (1987). "Coliphages as an indicator of faecal pollution in water, its relationship with indicator and pathogenic microorganisms.". *Water Res.*, Vol.21, No. 12, pp. 1473-1480.

CA RWQCB. (1999). "A Resolution Amending the Water Quality Control Plan for the Santa Ana River Basin To Establish a Total Maximum Daily Load for Fecal Coliform Bacteria in Newport Bay". Resolution 99-10, 1999.

Cabelli, V. J. (1983). "Public health and water quality significance of viral diseases transmitted by drinking water and recreational water". In: *Water Virology*, W.O. Grabow (ed.); Water Sci. Tech. 15(5): 1-15.

Champsaur, H., Questiaux, E., Prevot, J., Henry-Amar, M., Goldszmidt, D., Bourjouane, M., and Bach, C. (1983). "Rotavirus Carriage, Asymptomatic Infection and Disease in the First Two Years of Life. I. Virus Shedding" *The Journal of Infectious Diseases*, Vol. 149, No.5, May 1984 pp. 667-674.

County of Orange Public Facilities Resources Department. (1997). "Plans and Specifications for the Construction of Unit III Upper Newport Bay Sediment Control and Enhancement Project", funded by the State of California.

Dutka, B. J., Shaarawi, A. E., Martins, M. T., and Sanchez, P. S. (1987). "North and South American Studies on the Potential of Coliphage as a Water Quality Indicator". *Water Research*, Vol. 21, No. 9, pp. 1127-1134.

Eisenberg, J. N., Seto, E. Y. W., Olivieri, A., and Spear, R. C. (1996). "Quantifying Water Pathogen Risk in an Epidemiological Framework". *Risk Analysis*, Vol. 16, No.4, pp.549-563.

Eisenberg, J. N. S., Seto, E. Y. W., Colford, J. M., Jr., Olivieri, A., and Spear, R. C. (1998). "An Analysis of the Milwaukee Cryptosporidiosis Outbreak Based on a Dynamic Model of the Infection Process" *Epidemiology*. Vol. 9, No. 3.

EOA Inc. (1995). "Mamala Bay Study Infectious Disease Public Health Risk Assessment". Prepared for the Mamala Bay Study Commission.

EOA Inc., Resource Management Associates, and University of California at Berkeley School of Public Health. (2001). "Public Health Risk Assessment for the Newport Bay Watershed: Recreational Contact and Microbiological Risk". Prepared by J.A. Soller, S. Salvia, A.W. Olivieri, K. Kerr, R. Goebel, J.F. DeGeorge, S. Ginsberg, and J. Eisenberg for the Newport Bay Watershed Committee and the Santa Ana RWQCB.

EOA Inc., Systech Engineering, BioVir Laboratories, and U.C. Berkeley School of Public Health. (1996). "City of Stockton NPDES Studies Health Risk Assessment". Prepared for the City of Stockton.

EOA Inc., and U.C. Berkeley School of Public Health. (1995). "Microbial Risk Assessment for Reclaimed Water". Prepared for the Irvine Ranch Water District and the National Water Resource Association.

EOA Inc., and U.C. Berkeley School of Public Health. (1999). "Evaluation of Pathogen Risk Assessment Framework". Prepared by J.A. Soller, J.N. Eisenberg, and A.W. Olivieri for ILSI Risk Science Institute.

Fankhauser, R. L., Noel, J. S., Monroe, S. S., Ando, T., and Glass, R. I. (1998). "Molecular epidemiology of "Norwalk -like viruses" in outbreaks of gastroenteritis in the United States". *Journal of Infectious Diseases*, Vol. 178, No. 6, pp. 1571-1578.

Fleisher, J. M. (1991). "A Reanalysis of Data Supporting U.S. Federal Bacteriological Water Quality Criteria Governing Marine Recreational Waters". *Water Pollution Control Federation Research Journal*, Vol. 63, No. 3, pp. 259-265.

Fleisher, J. M., Jones, F., Kay, D., Stanwell-Smith, R., Wyer, M. D., and Morano, R. (1993). "Water and Non-water Related Risk Factors for Gastroenteritis Among Bathers Exposed to Sewage-Contaminated Marine Waters". *International Journal of Epidemiology*, Vol. 22, No. 4, pp. 698-708.

Flewett, T. H., and Woode, G. N. (1978). "The Rotaviruses: Brief Review" *Archives of Virology* 57, 1-23.

Fujioka, R. S., Hashimoto, H. H., Siwak, E. B., and Young, R. H. (1981). "Effect of Sunlight on Survival of Indicator Bacterial in Seawater". *Applied and Environmental Microbiology*, Vol. 41, No. 3, pp. 690-696.

Fujioka, R. S., Loh, P. C., and Lau, L. S. (1980). "Survival of Human Enteroviruses in the Hawaiian Ocean Environment: Evidence for Virus-Inactivating Microorganisms" *Applied and Environmental Microbiology*, pp. 1105-1110.

Gantzer, C., Gillerman, L., Kuznetsov, M., and Oron, G. (2000). "Adsorption and survival of faecal coliforms, somatic coliphages and F-specific RNA phages in soil irrigated with wastewater". 1st World Water Congress of the International Water Association, Health Related Water Microbiology, July 3-7, 2000, Paris, France.

Gerba, C. P., Goyal, S. M., LaBelle, R. L., Cech, I., and Bodgan, G. F. (1979). "Failure of indicator bacteria to reflect the occurrence of enteroviruses in marine waters". *American Journal of Public Health*, Vol. 69, No. 11, pp. 1116-1119.

Gold, M., Bartlett, M., McGee, C., and Deets, G. (1992). "Pathogens and Indicators in Storm Drains Within the Santa Monica Bay Watershed" Santa Monica Restoration Project.

Grabow, W. O. K. (2001). "Bacteriophages: Update on application as models for viruses in water". *Water SA*, Vol 27, No.2, 251-268.

Griffin, D. W., Gibson, C. J., Lipp, E. K., Riley, K., Paul, J. H., and Rose, J. B. (1999). "Detection of Viral Pathogens by Reverse Transcriptase PCR and of Microbial Indicators by Standard Methods in the Canals of the Florida Keys". *Applied and Environmental Microbiology*, Vol. 65, No. 9, pp. 4118-4125.

Gurwith, M., Wenman, W., Hinde, D., Heltham, S., and Greenberg, H. (1981). "A Prospective Study of Rotavirus Infection in Infants and Young Children" *The Journal of Infectious Diseases*, Vol. 144., No.3, 218-224.

Haas, C. N., Anotai, J., and Engelbrecht, R. S. (1996). "Monte Carlo assessment of microbial risk associated with landfilling of fecal material". *Water Environment Research*, Vol. 68, pp. 1123-1131.

Haas, C. N., Rose, J. B., and Gerba, C. P. (1999). "Quantitative Microbial Risk Assessment". J.W. Wiley, Inc., 1999.

Havelaar, A. H. (1987). "Bacteriophages as model organisms in water treatment". *Microbiological Science*, Vol. 4, No. 12, pp. 362-364.

Havelaar, A. H., Furuse, K., and Hogeboom, W. M. (1986). "Bacteriophages and indicator bacteria in human and animal faeces". *Journal of Applied Bacteriology*, Vol. 60, No. 3, pp. 255-262.

Havelaar, A. H., Hogeboom, W. M., and Pot, R. (1984). "F specific RNA bacteriophages as model viruses in sewage: methodology and occurrence". *Wat. Sci. Tech.*, Vol. 17, Amsterdam, pp. 645-655.

Havelaar, A. H., Olphen, M. V., and Drost, Y. C. (1993). "F-Specific RNA Bacteriophages Are Adequate Model Organisms for Enteric Viruses in Fresh Water". *Applied and Environmental Microbiology*, Vol. 59, No. 9, pp. 2956-2962.

Hethcote, H. (1976). "Qualitative Analyses of Communicable Disease Models". *Mathematical Biosciences* Vol. 28, pp. 335-356.

Hurst, C. J., and Gerba, C. P. (1980). "Stability of Simian Rotavirus in Fresh and Estuarine Water" *Applied and Environmental Microbiology,* pp. 1-5.

Hydroscience. (1977). "Task Report, Special Water Quality Studies (PCP Task 317)". City of New York, 208 Report, Hazen & Sawyer, Managing Consultants N.Y., N.Y.

IAWPRC Study Group. (1991). "Review Paper-Bacteriophages as Model Viruses in Water Quality Control". *Water Research*, Vol. 25, No. 5, pp. 529-545.

International Life Sciences Institute - Risk Science Institute. (1996). "A Conceptual Framework to Assess the Risks of Human Disease Following Exposure to Pathogens". *Risk Analysis*, Vol 16, pp. 841-848.

Jagals, P., Grabow, W. O. K., and de Villiers, J. C. (1995). "Evaluation of Indicators for Assessment of Human and Animal Faecal Pollution of Surface Run-Off". *Water Science Technology* Vol. 31, No. 5-6, pp. 235-241.

Kapikian, A., Wyatt, R. G., Greenberg, H. B., Kalica, A. R., Kim, H. W., Brandt, C. D., Rodreigues, W. J., Parrott, R. H., and Chanock, R. M. (1980). "Approaches to Immunization of Infants and Young Children Against Gastroenteritis Due to Rotaviruses" *Reviews of Infectious Diseases*, Vol. 2, No. 3, 459-469.

Kay, D., Fleisher, J. M., Salmon, R. L., Jones, F., Wyer, M. D., Godfree, A. F., Zelenauch-Jacquotte, Z., and Shore, R. (1994). "Predicting Likelihood of Gastroenteritis from Sea Bathing: Results from Randomized Exposure". *The Lancet*, Vol. 344, pp. 905-909.

Koopman, J. S., Monto, A. S., and Longini, I. M. (1989). "The Tecumseh Study XVI: Family and Community Sources of Rotavirus Infection" *American Journal of Epidemiology*, Vol. 130, No.4, pp. 760-768.

Kott, Y., Rozi, N., Sperber, S., and Betzer, N. (1974). "Bacteriophages as viral pollution indicators". *Water Research*, Vol. 8, pp. 165-171.

Levine, W. C., and Stephenson, W. T. (1990). "Waterborne disease outbreaks, 1986-1988". *Morb. Mort. Wkly. Rpt.*, Vol.39/No.SS-1.

Mathworks Inc. (1998). "Using MATLAB". The Mathworks, Inc., 1998.

McLaughlin, M. R., and Rose, J. B. (2000). "Evaluation of the *Bacteroides fragilis* phage as an alternative indicator of sewage pollution". 1st World Water Congress of the International Water Association, Health Related Water Microbiology, July 3-7, 2000, Paris, France.

Mead, P. S., Slutsker, L., Dietz, V., McCaig, L. F., Bresee, J. S., Shapiro, C., Griffin, P. M., and Tauxe, R. V. (1999). "Food Related Illness and Death in the United States". *Emerging Infectious Diseases*, Vol. 5, No. 5, pp. 1-28.

Metcalf and Eddy. (1981). "Wastewater Engineering: Collection and Pumping of Wastewater". written and edited by G. Tchobanoglous, McGraw Hill publications.

Morinigo, M. A., Wheeler, D., and Berry, C. (1992). "Evaluation of different bacteriophage groups as faecal indicators in contaminated natural waters in Southern England". *Water Research*, Vol. 26, pp. 267-271.

National Research Council. (1983). "Risk Assessment in the Federal Government, Managing the Process". National Academy Press, Washington, D.C.

Palmeteer, G. A., Dutka, B. J., Janzen, E. M., Meissner, S. M., and Sakellaris, M. G. (1991). "Coliphage and Bacteriophage as Indicators of Recreational Water Quality". *Water Research*, Vol. 25, No. 3, pp. 355-357.

Paul, J. H., Rose, J. B., Jiang, S. C., London, P., Xhou, X., and Kellogg, C. A. (1997). "Coliphage and Indigenous Phage in Mamala Bay, Oahu, Hawaii". *Applied and Environmental Microbiology*, Vol. 63, No. 1, pp.133-138.

Raphael, R. A., Sattar, S. A., and Springthorpe, V. S. (1985). "Long-term survival of human rotavirus in raw and treated river water" *Can. J. Microbiol*, Vol. 31, No. 2, pp.124-128

Resource Management Associates, Inc. (1997). "Upper Newport Bay Final Model and GUI Development and Implementation Report", prepared for U.S. Army Corps of Engineers, Los Angeles District, October.

Resource Management Associates, Inc. (1998a). "Upper Newport Bay Numerical Model Development-Baseline Conditions Analysis", prepared for U.S. Army Corps of Engineers, Los Angeles District, July.

Resource Management Associates, Inc. (1998b). "Upper Newport Bay Salinity Study", prepared for U.S. Army Corps of Engineers, Los Angeles District, October.

Resource Management Associates, Inc. (1999). "Upper Newport Bay Alternative Analysis", prepared for U.S. Army Corps of Engineers, Los Angeles District, November.

Resource Management Associates, Inc. (2000). "Newport Bay Water Quality Model Documentation", prepared for California Regional Water Quality Control Board, Santa Ana Region, January.

Sinton, L. W., Finlay, R. K., and Lynch, P. A. (1999). "Sunlight Inactivation of Fecal Bacteriophages and Bacteria in Sewage Polluted Seawater". *Applied and Environmental Microbiology*, Vol. 65, No. 8, pp. 3605-3613.

Sobsey, M., Battigelli, D. A., Handzel, T. R., and Schwab, K. J. (1995). "Male-Specific Coliphages as Indicators of Viral Contamination of Drinking Water.", AWWA Research Foundation.

Stetler, R. E. (1984). "Coliphages as Indicators of Enteroviruses". *Applied and Environmental Microbiology*, Vol. 48, No. 3, pp. 668-670.

U.S. EPA. (1986). "Ambient Water Quality Criteria for Bacteria". EPA440/5-84-002, 1986.

U.S. EPA. (2000). "Draft Implementation Guidance for Ambient Water Quality Criteria for Bacteria - 1986.". EPA-823-D-00-001.

U.S. EPA. (2001). "Protocol for Developing Pathogen TMDLs". Office of Water, EPA 841-R-00-002.

WHO. (1999). "Health Based Monitoring of Recreational Waters: The Feasibility of a New Approach (The 'Annapolis Protocol').", World Health Organization- Sustainable Development and Healthy Environments, Geneva.

Wommack, K. E., Hill, R. T., Muller, T. A., and Colwell, R. R. (1996). "Effects of Sunlight on Bacteriophage Viability and Structure". *Applied and Environmental Microbiology*, Vol. 62, No. 4, pp. 1336-1341.

Yates, M. V., Anderson, M. A., Gerba, C. P., and Rose, J. B. (1997). "The Impact of Body-Contact Recreation in the Eastside Reservoir Project on 1) Pathogen Risk to Consumers and Recreators and 2) MTBE Contamination". Final Report to the Metropolitan Water District of Southern California.

Young, K. D., and Thackston, E. L. (1999). "Housing Density and Bacterial Loading in Urban Streams". *Journal of Environmental Engineering*, Vol. 125, No. 12, pp. 1177-1180.

# WERF Subscribers

The following agencies and corporations are WERF's charter stormwater research supporters:

## Stormwater Agency

**California**
Monterey, City of
Santa Rosa, City of

**Colorado**
Boulder, City of

**Georgia**
Griffin, City of

**Kansas**
Overland Park, City of

**Kentucky**
Louisville & Jefferson County Metropolitan Sewer District

**North Carolina**
Cary, Town of
Charlotte, City of/Mecklenburg County

**Pennsylvania**
Philadelphia, City of

## Corporate

BaySaver Inc.
Vortechnics Inc.
WWETCO, LLC

## State

Fresno Metropolitan Flood Control District, Calif.
Urban Drainage & Flood Control District, Colo.

## Wastewater Utility

**Alabama**
Montgomery Water Works & Sanitary Sewer Board

**Alaska**
Anchorage Water & Wastewater Utility

**Arizona**
Gila Resources
Glendale, City of, Utilities Department
Mesa, City of
Phoenix Water Services Department
Pima County Wastewater Management

**Arkansas**
Little Rock Wastewater Utility

**California**
Central Contra Costa Sanitary District
Contra Costa Water District
Crestline Sanitation District
Delta Diablo Sanitation District
Dublin San Ramon Services District
East Bay Municipal Utility District
El Dorado Irrigation District
Escondido, City of
Fairfield-Suisun Sewer District
Irvine Ranch Water District
Las Virgenes Municipal Water District
Lodi, City of
Los Angeles, City of
Los Angeles County, Sanitation Districts of
Orange County Sanitation District
Palo Alto, City of
Riverside, City of
Sacramento Regional County Sanitation District
San Diego Metropolitan Wastewater Department, City of
San Francisco, City & County of
San Jose, City of
Santa Barbara, City of
Santa Rosa, City of
South Bayside System Authority
South Orange County Wastewater Authority
Union Sanitary District

**Colorado**
Boulder, City of
Colorado Springs, City of
Littleton/Englewood Water Pollution Control Plant
Metro Wastewater Reclamation District, Denver

**Connecticut**
New Haven, City of, WPCA

**District of Columbia**
District of Columbia Water & Sewer Authority

**Florida**
Broward, County of
Fort Lauderdale, City of
Gainesville Regional Utilities
JEA
Kissimmee, City of, Department of Water Resources
Miami-Dade Water & Sewer Authority
Orange County Utilities Department
Orlando, City of
Reedy Creek Improvement District
Seminole County Environmental Services
St. Petersburg, City of
Stuart Public Utilities
Tallahassee, City of
Tampa, City of
West Palm Beach, City of

**Georgia**
Atlanta Wastewater Services
Augusta, City of
Clayton County Water Authority
Cobb County Water System
Columbus Water Works
Fulton County
Gwinnett County Department of Public Utilities
Macon Water Authority
Savannah, City of

**Hawaii**
Honolulu, City and County of

**Illinois**
American Bottoms Wastewater Treatment Plant
Dupage, County of
Greater Peoria Sanitary District
Metropolitan Water Reclamation District of Greater Chicago
Wheaton Sanitary District

**Indiana**
Fort Wayne, City of
Indianapolis, City of

**Iowa**
Ames, City of
Des Moines Metro Wastewater Reclamation Authority

**Kansas**
Johnson County Unified Wastewater Districts
Unified Government of Wyandotte County/Kansas City, City of

**Kentucky**
Louisville & Jefferson County Metropolitan Sewer District

**Louisiana**
Sewerage & Water Board of New Orleans

**Maine**
Bangor, City of

**Maryland**
Anne Arundel County Bureau of Utility Operations
Howard County Department of Public Works
Washington Suburban Sanitary Commission

**Massachusetts**
Boston Water & Sewer Commission
Massachusetts Water Resources Authority
Upper Blackstone Water Pollution Abatement District

**Michigan**
Ann Arbor, City of
Detroit, City of

Grand Rapids, City of
Holland Board of Public Works
Lansing, City of
Owosso Mid-Shiawassee County WWTP
Saginaw, City of
Wayne County Department of Environment
Wyoming, City of

### Missouri
Independence, City of
Kansas City Missouri Water Services Department
Little Blue Valley Sewer District
Metropolitan St. Louis Sewer District

### Nebraska
Lincoln Wastewater System

### Nevada
Henderson, City of

### New Jersey
Passaic Valley Sewerage Commissioners

### New York
New York City Department of Environmental Protection

### North Carolina
Cary, Town of
Charlotte/Mecklenburg Utility Department
Durham, City of
Metropolitan Sewerage District of Buncombe County
Orange Water & Sewer Authority

### Ohio
Akron, City of
Butler County Department of Environmental Services
Columbus, City of
Metropolitan Sewer District of Greater Cincinnati
Northeast Ohio Regional Sewer District

### Oklahoma
Tulsa, City of

### Oregon
Clean Water Services
Eugene/Springfield Water Pollution Control
Water Environment Services

### Pennsylvania
Philadelphia, City of
University Area Joint Authority, State College

### South Carolina
Charleston Commissioners of Public Works
Mount Pleasant Waterworks & Sewer Commission
Spartanburg Sanitary Sewer District

### Tennessee
Cleveland, City of
Knoxville Utilities Board
Murfreesboro Water & Sewer Department
Nashville Metro Water Services

### Texas
Amarillo, City of
Austin, City of
Dallas Water Utilities
Denton, City of
El Paso Water Utilities
Fort Worth, City of
Gulf Coast Waste Disposal Authority
Houston, City of
San Antonio Water System
Trinity River Authority

### Utah
Salt Lake City Corporation

### Virginia
Alexandria Sanitation Authority
Arlington, County of
Fairfax County Virginia
Hampton Roads Sanitation District
Henrico, County of
Hopewell Regional Wastewater Treatment Facility
Loudoun County Sanitation Authority
Prince William County Service Authority
Richmond, City of

### Washington
Edmonds, City of
Everett, City of
King County Department of Natural Resources
Seattle Public Utilities

### Wisconsin
Green Bay Metro Sewerage District
Madison Metropolitan Sewerage District
Milwaukee Metropolitan Sewerage District
Racine, City of
Sheboygan Regional Wastewater Treatment
Wausau Water Works

### Australia
Sydney Water Corporation

### Canada
Toronto, City of, Ontario

### Mexico
Servicios de Agua y Drenaje de Monterrey, I.P.D.

### United Kingdom
Yorkshire Water Services Limited

## Corporate

The ADVENT Group Inc.
Alan Plummer & Associates
Alpine Technology Inc.
Aquateam–Norwegian Water Technology Centre A/S
Black & Veatch
Boyle Engineering Corporation
BPR CSO
Brown & Caldwell
Burns & McDonnell
The Cadmus Group
Camp Dresser & McKee Inc.
Carollo Engineers Inc.
Carpenter Environmental Associates Inc.
CDS Technologies Inc.
Chemtrac Systems Inc.
CH2M HILL
Clancy Environmental Consultants Inc.
Damon S. Williams Associates, LLC
Earth Tech Inc.
Ecolab Water Care Services
EMA Inc.
The Eshelman Company Inc.
Finkbeiner, Pettis, & Strout Inc.
Frontier Geosciences Inc.
ftn Associates Inc.
Gannett Fleming Inc.
GE Betz
Golder Associates Inc.
Greeley and Hansen LLC
The HACH Company
Hazen & Sawyer, P.C.
HDR Engineering Inc.
HNTB Corporation
HydroQual Inc.
Insituform Technologies Inc.
Institute for Environmental Technology & Industry, Korea
Jacobson Helgoth Consultants Inc.
Jason Consultants Inc.
Jordan, Jones, & Goulding Inc.
KCI Technologies Inc.
Kelly & Weaver, P.C.
Kennedy/Jenks Consultants
Komline Sanderson Engineering Corporation
Lawler, Matusky & Skelly Engineers, LLP
Limno-Tech Inc.
Lombardo Associates Inc.
Malcolm Pirnie Inc.
McKim & Creed
MWH
Odor & Corrosion Technology Consultants Inc. (OCTC)
ONDEO Degremont Inc.
PA Government Services Inc.
Parametrix Inc.
Parsons
Post, Buckley, Schuh & Jernigan
The RETEC Group
R.M. Towill Corporation
Royce Technologies
Sear-Brown
SYNAGRO
Tetra Tech Inc.
Trojan Technologies Inc.
URS Corporation
USFilter NATC
Wade-Trim Inc.
Weston Solutions Inc.
Woodard & Curran
Woodruff & Howe Environmental Engineering Inc. (WHEE)
WRc/D&B, LLC

## Industry

American Electric Power
BP-Amoco Corporation
Chevron Research & Technology Company
The Coca-Cola Company
Dow Chemical Company
DuPont Company
Eastman Chemical Company
Eastman Kodak Company
Merck & Company Inc.
ONDEO Services
Procter & Gamble Company
Reliant Energy
Severn Trent Services Inc.
Shell Global Solutions
Thames Water Plc
United Water Services LLC

# WERF Product Order Form

As a benefit of joining the Water Environment Research Foundation, subscribers are entitled to receive one complimentary copy of all final reports and other products. Additional copies are available at cost (usually $10). To order your complimentary copy of a report, please write "free" in the unit price column. WERF keeps track of all orders. If the charge differs from what is shown here, we will call to confirm the total before processing.

Name ______________________ Title ______________________

Organization ______________________

Address ______________________

City ______________ State ______ Zip Code ______ Country ______

Phone ______________ Fax ______________ Email ______________

| Stock # | Product | Quantity | Unit Price | Total |
|---|---|---|---|---|
| | | | | |
| | | | | |
| | | | | |
| | | | | |
| | | | | |
| | | | | |
| | | | | |
| | | | | |
| | | | Postage & Handling | |
| | | | VA Residents Add 4.5% Sales Tax | |
| | | | Canadian Residents Add 7% GST | |
| | | | TOTAL | |

## Method of Payment:

(All orders must be prepaid.)

- ❑ Check or Money Order Enclosed
- ❑ Visa ❑ Mastercard ❑ American Express

Account No. ______________ Exp. Date ______

Signature ______________________

## Shipping & Handling:

| Amount of Order | United States | Canada & Mexico | All Others |
|---|---|---|---|
| Up to but not more than: | Add: | Add: | Add: |
| $20.00 | $5.00* | $8.00 | 50% of amount |
| 30.00 | 5.50 | 8.00 | 40% of amount |
| 40.00 | 6.00 | 8.00 | |
| 50.00 | 6.50 | 14.00 | |
| 60.00 | 7.00 | 14.00 | |
| 80.00 | 8.00 | 14.00 | |
| 100.00 | 10.00 | 21.00 | |
| 150.00 | 12.50 | 28.00 | |
| 200.00 | 15.00 | 35.00 | |
| More than $200.00 | Add 20% of order | Add 20% of order | |

*minimum amount for all orders

Note: Please make checks payable to the Water Environment Research Foundation.

## To Order (Subscribers Only):

Log on to www.werf.org and click on the "Product Catalog."

Phone: (703) 684-2470
Fax: (703) 299-0742.

WERF
Attn: Subscriber Services
601 Wythe Street
Alexandria, VA 22314-1994

## To Order (Non-Subscribers):

Non-subscribers may be able to order WERF publications either through WEF (www.wef.org) or IWA Publishing (www.iwapublishing.com). Visit WERF's website at www.werf.org for details.

IWA Publishing's authorised EU representative for General Product Safety Regulations is Diane D'Arras, 15 rue Duret, 75116 Paris, France, e-mail: safety@iwap.co.uk.

Printed and bound by CPI Group (UK) Ltd, Croydon, CR0 4YY
12/05/2026
02108913-0001